Covid-19 – How it made us feel

Life in lockdown during the coronavirus pandemic

Louise Usher et al.,

Copyright © 2012 Author Name

All rights reserved.

ISBN: 978-1-8380517-0-9

The authors of this book express opinions which are their own, valid thoughts and feelings. With sympathy for those who lost their lives and had a difficult time through lockdown and the virus, the authors voices remain authentic and this is a likeable account of thoughts and experiences at this time.

Some names have been removed or changed where appropriate.

Covid-19: How it made us feel

FOR ROSIE

a walking angel.

And for those who left us too soon.

CONTENTS

Acknowledgments — I

1	Chickpeas and the virus	1
2	The New Normal	9
3	Requiem in C# Major	13
4	Asthma Fears	15
5	Final Year Biomedical science	23
6	Lockdown (poem)	29
7	Locked down in Bosnia	34
8	Is this like the War?	39
9	The day the schools closed	46
10	Mothers day in the jewellery store	50
11	Missing	58
12	Dad's on the front line	61
13	All gone mad	64
14	Switching off the glitter ball	72
15	No rain no flowers	75
16	Afterword – How writing made us feel	136
17	Glossary	145

Covid-19: How it made us feel

ACKNOWLEDGMENTS

With thanks to the incredible authors who gave their words
and emotions

Foreword

All of us seem so different. Different lives, different countries, the same planet. Colour, belief and age see no barriers just now. We are all human beings, susceptible to the Covid-19 strain of Corona Virus.

How has this been for us as different humans? What are our experiences, thoughts and feelings? Will we feel the same as each other and live the same groundhog days – or have we had different journeys?

This time in history is unique. We all initially, thought we were invincible. Until the numbers showed every day just how easily this virus was spreading. We started to realise the very real risk of dying from it. Everyone knew someone who had it or knew someone who had sadly died from the virus and that motivated us to all follow the rules, stay indoors and tut at people who didn't seem to be following the social distancing advice.

Human lives everywhere have been affected by this global pandemic and all of us have seen a myriad of emotions evoked during the lockdown. A few of us decided to journal our experiences to share in this anthology and piecing the tales together you can see a common theme uniting the human race yet all of us with vastly different voices.

Then the positivity began to arrive towards the end of April 2020. People were beginning to realise the positivity of being home, the lack of worry for the usual things despite the concern for the basic right of survival.

One of the loveliest things is to walk without a watch. Life began to feel very different in many ways. Memes trickled their way through social media, speaking of 'high maintenance' ladies who regularly had eyelashes, eyebrows, nails and roots taken care of by a salon which were unable to open just now. Those greys got longer, google answered 'how to remove acrylics at home' and 'how to cut your fringe with nail scissors' and men began to shave their heads. Facebook and Instagram stories saw photos of men with almost bald heads and smiles in a submissive fashion, occasionally a shoulder shrug.

Before the pandemic, patients complained about the National Health Service (NHS) waiting times or the fact the clinics were always running late. Things never seemed good enough. Yet now we are taking to the streets every Thursday at 8 pm with fireworks, banging pots and pans as we 'clap for our carers'. People have never seemed so supportive of our NHS. My appointment in rheumatology has been cancelled. No problem. Most clinics have been cancelled and people have accepted it without a grumble.

Number ten downing street must surely be full of meetings and crisis talks in this global emergency. New schemes to help us have been rolled out 'at pace' the Chancellor tells us. My knowledge of politics is limited but from what I have to judge on in front of me, it looks like those guys making the decisions are doing okay. I won't suggest y congrats to the Government on Twitter. I can't fathom that debate. Some like it, not me.

Holidays were cancelled, people were missing their loved ones. A vacant embrace being replaced by words and encouragement of, 'when we get through this, things will be even better.' These thoughts were carefully shared on social media as if to reach out to others in the hope of some encouragement of one's bravery. But not once had I heard anger or frustration. rather more a clenched jaw and an internal air-punching, 'come on!'.

Cash wasn't allowed in shops, if possible. Contactless please. My purse was only opened these days to remove a card for internet shopping. A lonely, blue £5 note sat in there week after week. Usually, that wouldn't last longer than a round of Starbucks.

I missed Starbucks. Not the coffee so much; the drive, the being there, the buzz. The other humans.

Instagram was full of posts like these:

'Ok so yes, I miss the coffee shops. Do you?

I can't lie, I love packing up my mint green rucksack with

- pencil cases

- laptop

- chargers

- earphones

- journal

- notebook

- purse

- glasses

And getting in the car with the music up loud, driving to the coffee shop in the sunshine. Dancing in the seat, tapping the steering wheel, making those other drivers smile with your contagiously happy mood.

Oh, happy day.

But we can't.

And that's ok.

We have all adjusted, shifted things. Lots of comments like this are being written about life changing. It's so different.

We will look back on this time and say, "do you remember when…?"

Fondly probably. Thoughts of the times when we had time to make coffee at home, how much money we saved, we could work on the laptop with messy hair, last night's mascara still trying to escape, wearing comfy clothes and a snoring shihtzu next to us.

The clock ticks in my living room just now and I've decided to make another cuppa then get to my exciting edits. This is nice.

But, @Starbucks you're perfectly safe.

We can live without certain things. It won't stop us missing them, wanting them, remembering 'times when…' and that's perfectly okay.

Isn't it?

Chickpeas and the virus

Nicola Brewster

How to make filling dinners using chickpeas?

What is the total population of China?

How do I make homemade hand sanitiser?

What is the normal body temperature?

and Can I eat sprouting potatoes?

are just some of the Google searches I've done in the last few weeks. Information I would never usually have needed to know but all are now things that I have extensive knowledge of since COVID 19 arrived.

I had never considered the possibility of actually opening the 'emotional support' tin of chickpeas that have been in the cupboard since around 2002. They have no real purpose other than to just be there to reassure me that there'll always be something to eat no matter what happens even though I have no intention of ever needing to put them in either mine or my family's mouths. It's the minimalist version of doomsday prepping if you like.

Covid-19: How it made us feel

In the early days of the virus, I was sure it was a storm in a teacup, no need to worry, certainly no need to be stocking the garage, loft and shed to the rafters with tinned fruit and tuna or buying dozens of toilet rolls.

My main concern was keeping perspective for my family. This is a virus, not a tsunami of plague which will end the world and all of humanity. Reminding them that the initial numbers of those infected in China were very small in comparison to the total population of the country, astounding them with my impressive totalitarian states expertise and how the country could be completely closed with no one leaving their homes and how the Chinese government had total power. I fielded questions about other viruses that had spread worldwide, SARS, bird flu and swine flu and their effect on us both as a country and personally and I eye-rolled more than I've ever eye-rolled before at the constant screenshot updates I received from Whuflu.com. All in all, though I thought I did a great job at providing level headed information and perspective without even the slightest hint of concern, panic or hysteria.

And then other countries started closing their borders, the world went a little crazier, the level of panic and internal scaremongering within my household rose to epic proportions and I wondered whether I did have a little more to worry about than I first thought. No matter though, we had plenty of toilet roll, we had 2 bags of pasta and most importantly we still had the chickpeas!

Videos of army tanks driving down the motorway, shops being looted and police in riot gear started making their way into my inbox and while I could see that the tanks were on the wrong side of the road so couldn't possibly be coming to London, and that the looting videos were dated and very obviously not recent, it didn't stop the panicked conversations that London and the whole of the UK was about to be locked down. I received pictures of an alleged letter from someone in a position of extreme authority stating that the army was being drafted in to help with keeping the streets calm and no matter what I said and how much of the obvious I stated I wasn't able to bring reality or perspective to those around me. In amongst the videos, links to newspaper article and screenshots, messages started coming informing me that 'my friend's mums, brothers' girlfriend has it what if I get it now' and 'my friend knows someone whose mum works in a hospital and has treated someone with it I'm going to get it now.' It wasn't too difficult to find ways of keeping my household calm in those kinds of moments, reminding them that the standard advice was to wash hands very often while singing Happy birthday twice over (and no it doesn't have to be sung at full volume so the neighbours can hear), and not coughing and spluttering everywhere would be enough to keep them out of harm's way and anyway these people are not even in the same county as us so the chances of contracting COVID – 19 from them was all but impossible, but the growing panic and concern was reaching its peak.

And then the schools closed their doors, pubs closed, restaurants closed, theatres, cinemas and libraries closed and it all got a lot more serious and the world got a lot crazier.

Photographs of supermarkets with empty shelves started appearing and long queues forming outside as people panicked and bought everything they could lay their hands on. I needed to go shopping, it was our usual day and I was resolute in that I certainly wasn't going to buy anything more than normal! Once in Tesco and seeing the empty fruit and veg racks, the lack of milk, meat, store cupboard food, cereals, crisps, cleaning supplies and bathroom bits it became clear that the week's meals were going to be interesting! It was fine though, I told myself, we had 2 toilet rolls, half a bag of pasta and those chickpeas were standing firm in the cupboard.

We ate the duck breast. It may look similar to chicken if you look from a distance and squint but it's nowhere near as versatile. We ate the cubed lamb but it isn't the new minced beef and the little piece of fish I managed to get my hands on and served with a selection of past their best vegetables wasn't my finest culinary hour. The rest of the week went downhill from there. We had food however and tummies that were full enough.

Then came the announcement; the country was closing down. All shops not selling essential items needed to close, working from home became a necessity unless an impossibility, shopping for essentials as infrequently as possible, leaving the house once a day for a local walk, run, or cycle and travel only when it is imperative. The advice was clear; stay in stay safe.

I'm trying to do just that although being in lockdown is odd. No hours idled away walking around the shops, no 'shall we pop out for a quick pint?', no friends or family randomly knocking at the door. The shopping trip to Tesco becoming the highlight of the week is pretty sad and by the looks of it, we won't be doing that together anymore as regulations get tougher. It's become more of a trolley dash than a shopping trip anyway. Social distancing means queuing outside the shop is the new normal, standing in hazard boxes to pay and wondering whether there'll be a stampede when the lucky winner of the shelf stacker short straw drags the toilet roll delivery out to the shop floor.

What I am loving, though, is seeing people going by the window suitably attired for their purpose...

-I'm wearing activewear so I'm on my daily exercise and am walking around the block,

-I have an empty shopping bag so make no mistake, I'm on my essential shopping trip and nothing more,

-I'm wearing a 3-piece business suit and carrying a briefcase or hi-vis jacket, grubby trousers and hard hat while carrying a tool belt so I definitely can't work from home and that's why I'm outside of isolation perimeters.

I'm trying to keep calm and carry on in true stoic fashion! Staying indoors means that proper housework is happening, those things I know I should do but never get around to, washing the skirting boards, cleaning the wardrobe door mirror (that was a mistake, seeing myself dust free and unblurred was certainly an eye-opener) and wiping down the blinds slat by slat. Cleaning the windows whilst dialled into a webinar may also have occurred.

Others around me are not quite so stoical bordering on the hysterical would be more appropriate. I am finding it increasingly tiresome having a laser temperature gauge waved at my forehead every time I clear my throat, and when a mouthful of tea went down the wrong hole and made me cough for what was perceived as too long, hazmat suits were already in the Amazon shopping basket before I'd caught my breath.

There's no doubt it's a worrying time, sadly the numbers are growing rapidly and plenty of people seem to have suddenly become armchair epidemiology experts. In reality, all we can do is listen to the advice, do as we're told and sit tight, wait it out and deal with the new normal that will emerge once this is all over

We may have had to cancel our holiday of a lifetime driving route 66, my sons GCSE exams may have been cancelled, it may be another year before I see my family from overseas and my partners 40th birthday plans may well be scuppered but so far, my loved ones and I are safe, well and happy and we still have the trusty tin of chickpeas in the cupboard.

The New Normal

Joseph Arnaud

Viral Pandemic Lockdowns trigger my anxiety; who knew? I am one of the lucky ones: I can work from home, my wages will be paid, I have plenty of friends who I already only see online, I even have a plethora of indoor hobbies and well-stocked 'projected table'. None of that stills the constant knot of tension in my stomach. I teach biology; at the beginning of the year I was teaching about pandemics, now I am living through one. Watching bland statistics on PowerPoint slides being brought into vivid focus by reality was horrible. Like watching a movie when you'd already read the book, being unsurprised by the plot twists but angry that the director wasn't doing something about the needlessly depressing bits. Still, I watched the parade of fake news and rumours of half measure and blind optimism. It made me laugh; it makes me cry. I had to switch it off and switch off.

The tension gets worse as the problems of friends and family accumulate, and I am stuck on the far end of a digital link, unable to help. A friend who loses a job, a father who can't get paracetamol for his arthritis, a sister caught backpacking in the middle of Columbia... The dreams start, the vivid near nightmares that stress always triggers in me: trapped, trying to prevent a nuclear meltdown; facing down a capricious Roman emperor. You don't have to be Freud or Jung to figure out what dreaming of being at the mercy of uncontrollable forces signifies. Yet life goes on.

Work is... strange. Teaching remotely via a chat room and setting online tasks. We always used to joke about teaching being easier without all the students getting in the way. The jury's out on that, but it's for sure and certain a lot less fun. The lockdown has stolen the two best things about my job, working with students face to face and the moment at the end-of-year when I say a few final words to those heading off to university. A week of remote assessment, chat room lectures and emails start new dreams; dreams of me standing in a classroom haranguing them all for missing homework and late assignments - more nostalgia than a nightmare.

Things go up and down. My Dad gets his paracetamol, my sister makes it home. The knot eases a little. I end up in a long queue with a woman with a cough and I hope I was socially distant enough. It tightens up again. I distract myself by baking my own bread, the first batch is doughy but delicious; the next is better. It has the added virtue of controlling how much of it I can eat. I've gone from moderately active to static and I can't afford the calories. I read, I teach, I listen. I am singularly failing to take advantage of the one silver lining of lockdown: time to 'get some writing done'.

Like, everyone, I strive to find that happy balance between stress and ennui, where my creativity lies. At first, I fail; too unsettled, still too anxious. I shop for Hoovers instead, as staying in generates more detritus than the sad wheezing thing that served for an emptier house can handle. I bake some more and run out of bread flour, regretfully I am forced to follow Marie-Antoinette's advice. I talk on the phone more, talk, not text! Slowly routine emerges from chaos, giving a sense of stability. Staring out the window towards the end of the second week of the lockdown, I start to feel it again: that spark, that incentive to write, I know I've finally begun to adapt; as I shape myself around the new normal.

Covid-19: How it made us feel

Requiem in C# major

Lulu Gunter

fermata

[fəːˈmɑːtə]

NOUN

to stay or stop, to hold or pause. Indefinitely.

We pause, at the mercy of the conductor's baton waiting with bated breath before plunging into the next glorious phrase of music.

The conductor is faceless, nameless, shapeless, shifting, undefined and undefinable. Wielding control so fierce and omnipotent capable of unprecedented dramatic crescendos and then teasingly moving to, diminuendo, legato, rallentando to rest.

So this musical landscape, playing out its macabre melody: will it reach its tremulous Fine or ironically be repeated over and over and over da capo with infinite starts?

We perform, but now with heightened senses. Vibrant colours jump out to startle. Sharpened vision: focused I see with new formed clarity.

I pick up scents on the breeze: home baking, freshly mowed grass, spring buds reminiscent of childhood, friends, home, family.

And I hear And I hear And I listen, I really listen revelling in the awakening of this rebirth, searching the depths and scope of this - the soundtrack of life.

Asthma Fear During Covid 19

Anonymous

During the start of 2020, a worldwide pandemic was going to hit us all, with devastating consequences to the human race that would go down in history. It's not something you read about, but we have lived it, breathed it, for some harder than others, and worst of all for some have paid the ultimate price with their life.

I have been an asthmatic since I wore short trousers but learned to cope and adapt to be in charge of it and not the other way around. The Covid-19 virus that is here and now is something else. If you are asthmatic it has a whole new meaning.

At the start of this new terror, the Governments of various countries told people to stay at home. If you are over a certain age, have underlying health conditions such as asthma or autoimmune issues you are to self-isolate for the next twelve weeks. We became known as the twelve weekers.

It was acknowledged that this group of people had a much higher risk of more severe illness if exposed to Covid-19. This new virus kills indiscriminately, it has no boundaries, colour, creed, age or where you are in life. It's like a horror movie that we are all living in or the latest games console game for you to play and survive to the end.

Self-isolation sounds easy but believe me it's not. This new invisible killer which has currently cost many thousands of lives in the UK, and more in the world over, is frightening, to say the least. Now having asthma, you are at further risk than the normal man in the street. This is because this new virus attacks the respiratory system, and people are ending up in ICU on ventilators – fighting for breath and their life. Because this virus has spread so quickly, the NHS has been stretched to provide enough ventilators for people with this virus. Big companies are now stopping their production of the normal industry to design, manufacture new ventilators and air supply systems for the NHS etc. The likes of Dyson have stopped sucking and are now blowing air, Mercedes and McLaren F1 teams are producing parts for Medical kit instead of super-fast cars, the manufacturing industry has gone mad.

Having asthma, I understand what it's like to have an attack and strain your body for that air you need, to exhale all the air in your lungs till they feel like an empty balloon. Your lungs feel like you have inhaled napalm and the burning and pain is beyond what most people can imagine. When you are unwell with asthma, you become very reliant on the blue best friend that asthmatics carry around. People notice this little blue bit of plastic but most take no notice, not realising the importance of this paramedic in a bottle. Once administered, it's like instant electric to your lungs, your body responds and the haze you were subjected to goes as quick as it arrives. Relief arrives like a surfer's wave.

You may feel a rush to the head, a little dizzy, probably much the same as if you had smoked a joint; I imagine. If your asthma is bad enough, you may also use a nebuliser, this again becomes very important and life-changing. The liquid is like mixed Red Bull and Vodka. You break the vial, pour it into the machine and sit back with the mask on, inhaling life back in yourself. It will take a few minutes for you to consume the opaque contents which have been vaporised, but afterwards, life is calm for a while. It will open your airways, calm your burning lungs and give you a feeling of relief.

This is normal life for an asthmatic. We accept it, but we don't like it. But the rules are changing; there is a new sheriff in town called Covid-19. If you catch this virus, may God help you. On a bad day, an asthmatic can hear themselves breathing, and it consumes your thoughts. You will struggle to talk, eat or drink – you will just breathe until you are comfortable enough to do something else. If someone talks to you or heaven forbid asks you a question, you do not have the air to answer, and you will naturally try to answer even if it makes you worse. You will resort to using hand signals for things you need, the speed of which may quicken on the urgency of what you need. If you have been coughing all day and fighting asthma you will be tired and exhausted by bedtime.

When you lie down for some rest and hopefully relief, this is when asthma is at its most cruel. It fights back with a vengeance and the coughing seams constant and never-ending. You lie down and if you do manage to get some sleep – the thought will cross your mind many times – will I wake up in the morning?

You become short of breath, go a little hazy, because you are knackered you may think this is it…..sod it I have had enough and give in. The body being the marvellous invention it is, re-energises and as quick as you feel faint, you regain air and continue to live.

You sleep upright, with more pillows than before, a drink on the bedside table but most important of all is that blue inhaler is visible at all times and to hand. It could and does save your life through the day and night. Just to see it and then have it at arm's length is comforting.

Because of Covid-19 people are told to stay at home to stop spreading the virus. Businesses are putting staff on furlough; a bit like gardening leave. Shops, pubs and cinemas are closed until they are told they can open again to prevent the spread of the virus. No public gatherings, no meetings, no shaking of hands – the simplest gesture a gentleman can do to say hi.

No travel, no flights, no trains, no non-essential anything pretty much. All of which plays on our mental state. No one wants to catch this virus or pass it unknowingly to others or your loved ones. When you go to the shops for essential food shopping, you are to observe "social distancing", 2m apart from anyone else.

Wash your hands you are told, observe the social

distancing and all of which will save lives. However, people still need to eat, and function as normal as possible. As you stand in the queue for your local supermarket for essential food, you don't know if the person before you or after you has the virus, as some people do not have symptoms which is frightening. It's a risk.

Supermarkets now have markings every two meters apart, called hazard boxes. A direction arrows of which way to walk around the store, bit as Ikea has done for many years so you don't get lost. You feel like a sheep at a cattle market, but all the while its helping people stay alive, you want to follow the lines and arrows of the rule. In the supermarket today, I coughed, because of my asthma, I managed to clear the isle in seconds; everyone left. The fact is, I waved my inhaler around like a white flag but no one noticed. The fear of this virus is quite something to behold, feel and live through.

Especially if you are asthmatic or have any underlying health conditions. I believe as an asthmatic, if I was to catch this virus, I would not rate my chances of seeing my next birthday or Christmas. This is frightening for all, but on another level for some. History is being made, our mental state is being stretched and tried, but just to write this is a privilege as so many have left this earth plane because of this virus.

The world will never be the same ever again. I salute all the NHS, Dr's, Nurses, support staff, Pharmacists, key workers – people in food shops, delivery drivers etc – general people who are doing their job to keep us all going with necessities. These people are putting themselves at risk for others. Many have already died doing what to them is just their job. They've not to be forgotten I'm sure and they will all go down in history as hero's every one of them.

COVID-19: A Final Year Biomedical Science student prospective.

Megan Kerr

We had months left of university. Months with our friends and months to prepare for our future. This turned into days, then just hours before students hastily returned to their homes around the world before the lockdown grounded us all. We had our last lecture together and did not even know.

I had been at a Trypanosomiasis and Leishmaniasis conference in Granada Spain in the week leading up to the lockdown. The most pressing topic being coronavirus. Many people couldn't make it as they had already gone into lockdown. I remember sensing how real it was as staff from other universities were discussing how they would continue when (not if) a lockdown happened in the UK.

I got home on the 12th of March. It was two days later when Spain announced their lockdown. The following day we all had an email informing us our face-to-face lectures had been cancelled. Everyone was in shock. It's something many of us could see coming but at the same time hoped that it was a decision that would not have to be made. Then only 7 days later the UK announced its lockdown. The university was closing its doors.

It was like the wind had been knocked out of us. A light went out and no one knew what to think, or what to do. As the days went by we were inundated with emails, leaving us with more questions than answers. But the truth is no one had the answers. This had not happened in our lifetime; it is not something that has needed to be considered in a risk assessment or be prepared for. It's the sort of thing you hear about in movies.

As university students we are not the most affected by this disease. We tried not to feel selfishly negative. Thousands of people are dying every day, or risking their lives to save others. But we are grieving, a different kind of loss.

Over the last few years all of us students have had something to keep us motivated, to push us through the many deadlines, to overcome the stress and tears associated with studying a challenging degree. Whether it was taking that picture, submitting our final dissertation just like ones we had seen from the students before us, the celebratory drinks after our final exams, the graduation ceremony we had dreamed of for over three years, or just the thought of a restful night sleep.

All of which has come to a halt.

We have all submitted our dissertation at the click of a button behind a computer screen, our exam is to be taken online, graduation postponed indefinitely, our final goodbye over a video call. Some will be friends for life, but many we will never see again.

We are still working in less than ideal circumstances, trying to maintain motivation to continue our remaining assignments whilst worrying about our family, our health and our future. Some students are now working on the frontline, helping fight COVID-19, putting themselves at risk before they have finished university. As medical science students we are all aware about how serious this pandemic is. We have studied spread of diseases and the impact of disease on the body. I have found myself getting distracted and researching coronavirus and not working on my dissertation. I have been tracking the figures each day, creating my own graphs and comparing our results to other countries.

Before all this happened, I had a strange fear of talking on the phone or of video calls. It is something I would avoid at all costs but as course representative for Biomedical Science I have been in weekly meetings with fellow students and staff to ensure our voice is heard. To ensure we can achieve the grades we have been working so hard for.

As a disabled student my time at university hasn't been easy and I am not the same person I was when I started my degree. The last three years have been the best and some of the worst times of my life. I have overcome two lots of surgery, hospital admissions, declining health, family loss and care responsibilities for my disabled Dad and Grandad. With half of my family considered 'extremely vulnerable' and having to shield for at least 12 weeks the threat of this virus is very real for us. My dad can't leave his bed due to MS and is often admitted to hospital due to severe infections. His body cannot even fight a simple UTI. If he contracted the virus there is a very good chance he will never come home.

Being a student has completely changed my life. For the first time, I have not felt like I have had to prove myself to anyone. Staff at the university know what I am capable of, they see how determined I am and will do anything they can to break down the barriers I face, not put more in my way. The confidence this have given me has led to many fantastic opportunities. Instead of being told I can't do something, I am now asked 'what can we do to support you in doing that?' This has made me believe in myself and my abilities again. I am now preparing to move over 250 miles away from home to start a Master by Research in Biology at the University of York.

One thing this pandemic has taught me is that I need to stop, and breathe. I cannot allow myself to return to the normal I had before lockdown. That normal was making me unwell and I did not give myself the time I needed to focus on myself.

My memories of university are not going anywhere. I will remember them for the rest of my life. It is not the way I wanted to finish my degree. It's important to stay positive during this time and stick together. I have seen a community of students, who have never been so far apart, come together and support each other with admiration. Everyone's COVID-19 story will be different, and we will all have our own unique challenges, but the important thing is, we will get through this.

We will be successful.

We will have a great story to tell.

We will always be the ones who graduated university during a global pandemic.

This is just the beginning of our story.

Megan can be found on Linked in:
https://www.linkedin.com/in/megancherisekerr

Lockdown

A Poem by Ronnie

With the world on lockdown, you got some time to think, no time for socializing, nowhere to get a drink.

People going out and meeting, playing in the streets, whilst others risk their lives at work, people dying at their feet.

For me it is quite easy, I'm paralyzed in my bed, I make my freedom, with ideas deep within my head.

I see so many pictures of nurses all in tears, begging us to stay inside, they tell us of their fears.

But it seems that most don't care, they take no heed of the nurse's plight, do they have to lose someone before they do what's right?

Children dying all alone, no parent to hold their hand. Do you want this for your kids? Why don't you understand?

Stay home, keep your distance, do it for your loved ones and your friends, put a halt to going out and help this Virus end.

Yes, we all want a hug it's awful, boring on your own. Use the safety of the internet, use your mobile phone.

Only go out when you must, just shop to get your dinner, don't buy more than you need, don't be a hoarding sinner.

If you can help your fellow man or a neighbour who's in need, do it now, be helpful, it is your civic deed.

So, whilst I'm lying in my bed, please keep your kids indoors, go back to an earlier time, make them do some chores.

Be safe, wash your hands and love people from afar. Stay home to win this, no going in the car.

So, hope you all hear my plea, and everyone you love makes it through, I'll see you on the other side of this and remember I love you too.

Note from Louise: Ronnie is a disabled friend with a big heart. He posted a video on Facebook,

"Right, I'm not sure how to do this but I just feel strongly and I wanna say some things.

I know we've been having a laugh and I've been putting

stupid pictures up and silly jokes and all the other stuff but I'd like to say a huge thank you to me sisters, me friends, both in life, on Facebook, around the world who have helped carry me through the crap time I've been having.

Being paralysed ain't been much fun. But you lot have made it bearable. With all this virus going around, I just want you all to know, I love you all dearly.

Stay safe, follow the guidelines.

Kids out there, I know it's boring and you're gonna hate it, you don't wanna do your work at home, your schoolwork is boring....tough! It's just tough shit. People are dying and it's your job, my job, everyone's job to try and prevent the spread of this virus.

I'm old, I'm infirm, vulnerable or whatever you wanna call it; but I don't fancy dying just yet. I've still got so much mayhem to spread. So, just think about your loved ones, think about the most important people in your lives. And *you* being safe makes them safe. So, just be careful.

Thank you once again to everyone, I can't name yous all as it's too long and you'll be pissed off and bored! But I love you all and I'm so grateful for everything. Be safe, keep your distances, wash your hands and love each other"

Locked Down in Bosnia and Herzegovina.

David Bailey MBE

I'm David; British, born in London and now a resident of Bosnia and Herzegovina.

It's the start of April and here in our village, in the rural north of the country, life is still slowly changing as we get to grips with the effects of Coronavirus.

As I write it's been some 5 weeks or so since it became

apparent to this tiny country in the Western Balkans that the population would not be escaping the global pandemic.

I say became apparent, as Bosnians have an opinion of themselves, that they're unique and different from other cultures. That they might escape illnesses etc that more developed countries have.

COVID-19 arrived and people started to get infected.

Bosnia and Herzegovina is small: a population of around 3.8 million. Corruption is rife. Sores from the inter-ethnic conflict of the 1990s are still raw. 'Dysfunctional' sums up Bosnia and Herzegovina in a word.

Although I love living here dearly (as I have for nigh on 20 years), I did have doubts as to whether the country could handle a catastrophic event such as Coronavirus. So many ifs and buts went through my head. I mean you need to be well organised and proactive in dealing with large scale "disasters". You need resources as well.

Although the first few days were chaotic as the country reeled from the realization that COVID-19 had arrived, authorities, especially in the north of the country, gripped the situation and started to implement control measures.

From asking the population to self-regulate (which I think the population found difficult, as they did in the UK), to creating and enforcing measures to protect the health of the nation, through quarantine orders and a daily 10 curfew.

As the days went on, I found myself, although naturally worried about becoming infected, feeling more confident that the authorities had this situation under more control than bigger countries, such as Italy, Spain, France or Austria for example.

Of course, the country has knocked on many doors for material assistance, and thankfully resources that the country did not have in reserve, have started to trickle in.

The standard of Balkan medical staff is first class, and now with an ever-increasing toolbox available, these heroes will do whatever it takes to protect and assist their fellow citizens. Should I require help, I don't fear that my care would be of a lesser quality to that offered in London, Paris or Berlin.

I am self-isolating, together with my wife and her immediate family in a rural village. No neighbours on top of us and sufficient space to go for a walk without leaving the property. With our typically warm April weather to lift our spirits as well, we are blessed.

Rural Bosnians are well versed in self-sufficiency, so I worry not in that respect. Our only negative points are that we are not allowed to travel far, older people not at all (there are minor exceptions), and that the constant stream of news is almost only Coronavirus focussed. Something which I think might affect our mental health. People who live in apartments in cities might have mental health considerations higher on their radar than we do.

Watching how other countries struggle, makes us all here (well at least my wife and me) realise how lucky we are in comparison. There's still a long struggle ahead. Will I continue to escape infection? What waits around the corner?

Questions, questions.

As I finish these thoughts we have been told that the two Easter weekends, (one Catholic and the other Orthodox), will not be as traditional as they normally are.

Family members visiting from abroad will be required to quarantine at the border for 14 days. That precludes having the gatherings that are so much part of this countries Easter Celebrations.

My life has changed. Again.

What lies over the horizon, I have absolutely no idea.

David Bailey MBE can be found blogging at anenglishmaninthebalkans.com

and in 2019 Louise had the pleasure of taking a self-directed writer retreat to this tranquil part of the country which you can also visit as you can stay in this beautiful holiday apartment (traveltobosniaandherzegovina.com)

Is this like the war?

Tesco Worker

(anonymous)

I haven't lived through the war but blimey, I can only imagine something like this was seen in the war times. But they didn't have supermarkets then from what my Gran told me. It was a separate shop for the meat, the veg, the bread and stuff. I'm not even sure if they had things like rice and pasta back then. They must have had toilet roll I assume. Although Gran had this metal thing screwed into her toilet wall with greaseproof paper inside it. Mum told us we had to rub it together to try and make it softer and hope it might mop up some of us kids pee. Odd set up if you ask me.

I carry this guilt with me that when I used to sit on Gran's loo, I picked out the dried-up grout from between the lemon-yellow tiles and let it fall to the floor. Took my mind off how cold it was sitting on her toilet seat. It felt like porcelain, but I'm sure it was plastic.

A few days before the lockdown was announced, people started to panic buy. None of us inside the store understood what was going on and its kind of caught us by surprise. It was almost as if the public knew something before we did and I remember that first day, which seemed like forever ago now.

"There seems to be a shortage on the shelves of toilet rolls," the line manager told my friend who was stacking in that aisle. She went out to fill up and no sooner had she completed that, it needed doing again and we began to wonder what had happened which meant everyone needed toilet rolls.

"Look!" she told me during a stolen moment as she pointed to the aisle. People had their phones out, taking photos and sharing to social media, demonstrating that people were stockpiling loo rolls. It just made people think they also needed to come into the store and do the same and more people started to buy toilet rolls over the next couple of days. Pasta and rice were next and we began to wonder if people might start stocking up their underground bunkers or something.

We all rolled our eyes, metaphorically speaking, at this behaviour. We couldn't quite believe what people were doing as the whispers (which were doing the rounds between the staff) was that this was just like the flu which was in another country. Some of the staff followed the news and one guy who worked the deli (and seemed to know about these things) told us, "if you have an underlying health condition, you have to be careful but most people don't even need to go to the hospital for it. More people die of the flu. I don't understand what the fuss is about. It's in China anyway." And we *almost* took his word for it until word started to filter down from head office and then we all began to get a bit concerned. This was only a few days before the Prime Minister announced lockdown. Well, it wasn't lockdown for us of course.

The bosses told us to prepare for some crazy times ahead. We still weren't all that sure what they meant but the actions of the shoppers seemed to highlight to the rest of the British population that things were about to change but still we didn't expect what happened next.

Customers were taking things out of each other's trollies. If one noticed, there was a tug of war happening over the item. I saw my friend from the bread aisle run from her post out into the back, past all the posters on the tiled wall which remind you about customer service and COSHH and took to safety in the staff toilet area.

Following her at some speed, I caught up with her and put my hand on her shoulder as she turned to me. Her eyes were bloodshot and wet and soon tears ran down her cheeks.

"I can't deal with it, they are being horrible," she cried and I hugged her. Her body weight fell into my chest and I told her it would be okay,

"Just a few days of this and things will settle down, soon as they realise,"

"I know! If only they realised we were not going to run out of stock! If they would just give us a chance to catch up and put more stock out. There's no need for this," she said. She was right. But it didn't stop as people began to worry more.

Jokes flew around the internet.

'This virus doesn't cause diarrhoea.' People would reply with cry-laughing emojis. If only they could see inside the store and realise this wasn't a laughing matter at all. My friends at work would be in tears every day. The customers were frustrated and worried and it made them stressed and rude. They seemed to forget that we were trying to be helpful, trying to do our job. We were told in smaller groups that things were going to have to change in our store, and all Tesco's. The bosses were amazing actually with so much support for us all. They started handing out application forms, knowing we would need more staff and quickly. A few days before lockdown, we stopped being open 24 hours and the public hated that. Saturday morning, once people had started to get to know of people in hospitals in our country, there was an air of 'oh my goodness this is about to get very real,' and even while the grey shutters were down a massive queue extended along the front of the store and right up to the KFC next-door.

Social distancing wasn't invented then. Perhaps the length of the trollies might have saved a few people from contracting the virus as everyone in the queue had a trolley. Who knows how many people were walking around infecting people then, coming into our store, not washing their hands enough, and handing us cash.

As I was driving in to work that morning I saw someone driving his car through the car park with his phone in his hand, clearly filming the queue to put on the internet. He was alone in the car but talking and shaking his head in utter disbelief at the madness of everyone trying to get in and do their shopping. Yet he was also trying to do the same and all the other head shakers were too. They were all the same.

This was over three weeks ago now. Before lockdown and the social distancing. Before we allowed them to queue at a distance of two meters apart and let them in one by one. One in; one out. Keep your distance, stand behind the visor, use contactless payments. I wish they had set up a rule for trying to be nice to the staff who were trying to keep the country fed and in a good supply of food, toilet rolls, pasta and rice. Every day there were tears and yet we were putting ourselves at risk to do our job.

I began to worry about money. What if there was an issue? My husband is a barber and they were all told to stop work without any idea of what might happen with their income situation. He is self-employed. I think they are going to help the self-employed somehow. The chancellor said. But we don't know what that is yet. And we don't know how we pay our mortgage. Although they

We're saying we might be able to take a break in mortgage payments, called mortgage holidays. That sounds a little worrying but if we must, we must.

Then Millie went down with a temperature a week into

lockdown when it was all still mad at work. A proper big temperature. She was cutting her first teeth so it might have been that but the guidance was clear. Everyone must stay home for fourteen days. Oh crap.

The day the schools closed

Ben Odero

He told us that we would be safe. He told us that we had nothing to worry about. Even when our neighbours across the globe began to shut their doors. Back then, there were only whispers of a situation, far away from us. So far removed we didn't have to worry.

Right?

It was so easy for us to stick to our routine – to fill the pubs, the clubs, the parks. We didn't have a care in the world. The biggest commercial concern was what we'd get up to in the weekend. I have no excuse, I was the same. Between my love to find new books and the love of seeing new things, I wanted nothing more than to explore the rest of the world. See what else is out there for me, become someone new.

I remember the first day there was a rumour of work shutting down. We still didn't take it seriously.

"Who else is ready for some annual leave?" A colleague asked. "I needed a break anyway."

If only we knew.

Unknown to many, I had booked off time for my partner. For her birthday, late in the summer, when the sun would be high in the sky and the temperatures would soar as high as one's spirits would in the summer season, I would take her to Italy. She had never been, and quite frankly I had never been so excited to show her.

Italy closed its doors only days later.

I don't know many people, not at all. Knowing too many people causes too much stress and worry. How do you know who to check in on? Would they even want to talk after it being so long? What do you even say?

I make it a general practice to keep the important people close. As close as I can imagine. Now, it would be a godsend if I could meet a friend closer than two metres. And when I think about how comfortable our lives were before, I regret taking things for granted. Now, we live in an age where Facetime, Zoom, and Microsoft Teams reign. And while it isn't so bad, I wonder: could I have reached out to people more?

Now I promise you, lockdown isn't all bad. If anything, there are some great points. My family, who I love very much see me a lot more, and I do feel like I have more meaningful conversations with them. My dog, who very much loves to play every second of the day is overjoyed that I'm home. And that's one thing that has made me pause and think – while I am at home, keeping safe, some so many other people are not, by definition, safe. The healthcare workers, the retail workers. They have their backs against the wall, helping everyday people like me. They're making sure that we can live our lives as comfortably as we can, while they're at risk.

Doing something takes away the feeling of entrapment. These days, I wonder: what else can I do for the people who can't afford to stay home? The people who support their families by working? What else can I do? It's a question I ask myself every day. I can volunteer. I can do food runs for my community. There is so much that can be done, and I wonder: am I doing enough?

It's hard to put into perspective when there are so many others risking their lives so the world doesn't crawl to a standstill. But I know I will use this time to reach out to people both far and near.

The first thing I noticed on the day schools closed was how silent my work friends were. It was as if all the jokes about time off, all our worst fears had been rolled into one. We didn't know what to do with ourselves, but the message from our workplace was clear; go home.

Now, I've found that I'm going back to writing. It is a comforting thing for me – knowing that I have a blank document to work with. I think the one thing I would figure out to encourage others is to take the pen up yourselves. Tell everyone your stories. You're not in this alone. We are in this together, whether we are working from home or working outside. We have our own story to tell.

Mother's Day in the jewellery store

Jasmine Usher

The very first time I heard about this virus, I was on holiday in the beautiful sunny Caribbean side of Mexico. We enjoyed watching wildlife I hadn't seen before; monkeys, iguanas and coatis which looked like anteaters with long upright cat tails. The coatis were particularly cheeky, stealing what was left of the icy Piña colada from the side of our sunbeds. It seems an age ago now the coronavirus was nothing more than a mention that was soon forgotten about.

A few weeks went by and we began to hear 'Covid-19' more. I didn't see any signs of people being worried, we did not think we were a threat at this point since it was not in our country. After a little while, a few people in England started getting it but it all felt very well contained. Next came another new term, self-isolation. Meaning, if you show symptoms, stay away from other people.

Then came social distancing. It all started to feel a lot

more real.

I was working at a busy shopping centre not far from home, in a brightly lit jewellery shop. Serving our customers with pretty silver charms and rings, we need to be physically close to them, fitting their jewellery, near to their potential virus-containing bodies. Was I to worry?

Mother's Day was just a few days away. My twin brother and I shared the planning for our Mum. She had done a lot for us over the years and we loved to spoil her, especially as she sometimes isn't so well. I try not to worry about her. She is very strong, but I know sometimes things are not as easy for her as the painted smile would like to have us believe.

This was a typically busy time of year in our store, the ideal gift for mothers everywhere. Our usual management briefings began to change from talking about sales and customers service, to restrictions for our safety and lots of talk about hygiene and health and safety, taking care of everyone in store. I went for my lunch break to MnS café and saw the lady who usually makes coffee was pout purely on 'wiping' duty. Just wiping down the service counter with a J cloth and spray.

"It's my job all day long now," she replied to a customer

who asked if that was her role for the shift. Over the top? I wasn't sure.

Through the window to the left of our store, it was glass from floor to ceiling. In our view was the entrance to Sainsbury's. Customers leaving the store had trollies piled high. So much of what I could see was toilet roll in everyone's trolleys. Why? This didn't seem necessary, although, everyone seemed to go along with it because the shelves were empty as if people were expecting a plague. To begin with, I was very naive and unaware of the pandemonium this virus was causing. I thought it was just the bad flu everybody said it was, but as time went on it seemed like something else completely.

The news and social platforms were all filled with negative comments. Customers were talking about it and what they think, it's all we heard about for a while. Conversations everywhere were about the coronavirus and it made such a difference to the day, listening over and again about the virus. Shops became empty and people started getting more fearful. I felt nervous. What if one of the customers I was serving coughed near me and had the virus? What if I took it home? Senior management meetings were far more frequent than usual as whispers began to rustle through the store about our closing.

'Were you busy today?' Mum asked,

'yeah, really busy,'

'oh God I can't believe people are going out to buy jewellery.' Mum was shocked. She was already in quarantine following the advice she was given from the charity who deal with her chronic diseases. They posted on the website and Twitter to advise people like her. She was feeling unwell with a bad cough anyway. The hospital told her it was pneumonia.

With my contract ending on Mother's Day, management had said it looked very likely they would extend this contract, maybe even to a permanent contract. I was so pleased with this news! The contracted hours were perfect for me to fund the shiny, white new car I had just proudly purchased and the rest of the week would allow me to continue working on my online business. Life had been sailing downstream with ease recently. Things were good.

This all seemed a bit up in the air now, with my last shift on Friday and corona conversations echoing, I expected this to be my last shift ever. Shops were closing, people were losing their jobs, more new words occupied our vocabulary.

The following day I took a phone call from work announcing that the shop was going to be shut for the foreseeable future. I felt a slight relief, I didn't want to be working at a time like this, nor did I want to be putting my family at risk. A very odd feeling that I wasn't sure I should experience. The loss of one's job is often the absolute end of the world, so why did I feel so selfishly positive about it? My boss didn't mention on the phone about contracts; this was a very uncertain and confusing time for everyone. We were told to sit and wait patiently.

People started losing track of days, me too, but a few days or a week went by and I got told to expect another phone call. I felt completely unsure of what this would be about. My iPhone lit up and began with its ringtone and I nervously answered.

The Government had found another new word for us to add to our vocabulary; furloughed. This meant we, us, our jobs were being put on hold. So, we still had a job, we just weren't working there at the moment. Luckily, 80% of our wages could be covered by the Government so I continue to get paid. Knowing I could now totally isolate, get paid, catch up with life, I felt blessed. My thoughts relaxed a lot and I choose to look at fewer news articles, relieving the anxiety I might have otherwise faced.

I'm locking myself away from the world for now and that's all I know.

I can't help but count my blessings and feel I am extremely privileged at this time. Frightening as this is and tragic that so many are losing their lives, my personal feeling is one of gratitude. As much as this whole pandemic sounds completely awful, I've found myself in a very fortunate position: still employed, food in the cupboards, my health is good and I have a sudden wave of motivation. I feel this is the time to sort my life out, do what I love and make more content to put out online. After months of not feeling motivated, I finally found what I had been longing for. A different kind of feeling which was evoked by the lockdown, emotions of survival surfacing, renewing my self-belief and passions. So many ideas for plans I want to do, videos to make and photos to be posted. The world is in a worrying position just now but looking at the bigger picture, it's beautiful.

We are saving our globe, wildlife is returning to where it belongs and the planet is healing. I think we needed this. I am taking this great opportunity of time off to do everything that had been on my overly long to-do list for months. Ticking everything off one thing at a time and adding more to the bottom of the list just as fast, it's still never-ending but it's not overwhelming. I'm excited to be doing a lot and getting everything sorted. I have new ideas on ways to create the life of my dreams and travel to incredible places with the people I love, once this is all over. Even smaller plans like a pub lunch or casual drink with your friends could transition into a (not-so-casual, accidental) night out to remember. I won't take this for granted again. Once this is all over, it seems as if everyone is now desperate to strengthen those bonds they have with everyone and check up on loved ones more often. Let's hope that wish remains.

I am choosing to see the good in this situation as an amazing time to do things I've wanted to do for so long.

The whole street seems to think the same by the amount of bin bags out on the rubbish day. So many tidy houses by the end of this all. But I'm uncomfortable with the prices being paid by people at the mercy of this uninvited guest.

Who knows what will happen when restrictions begin to be lifted, will the economy be ok? Will I get my job back? Who knows? What I do know is that we will deal with it when it comes to it and make the best of the situation because things always turn out alright in the end. Mum says.

Jasmine can be found online as a keen influencer on

Instagram : https://www.instagram.com/justjasx/

Youtube: htttps://www.youtube.com/justjas

Pinterest: https://www.pinterest.co.uk/justjasx/

Another Poem from Ron

Missing simple things with the world on lockdown, will the Virus ever end, I'm missing all the simple things, like time spent with a friend.

Missing all my loved ones, they all seem far away, even though I hear from them every single day.

I read how people must try harder once this lockdown through, try to be nicer people, if only it were true.

Even those who I haven't seen, except for funerals and weddings are finding time to say hello, this Virus does my head in.

It won't take long for the tide to turn, for greed to rear its head, then it's back to how we all were, I'll see you when someone's dead.

We will talk of the Virus like it was a third world war, of rationing and lockdown when we cared for people more.

How we struggled with nothing, no pasta or bog roll, how shopping took for hours, so many people put on the dole.

How front-line staff were heroes, some paid the highest price, saving lives the best they could, we clapped for them; so nice.

But will we remember this time, once the Virus is done, will we be much nicer, when we're let back out into Sun.

Or will we turn back into the selfish sods like we were before, will we step over the needy, or people sleeping on the floor.

Let's hope we learn the important things, don't live inside a bank, its people who were there for you - you lot I'd like to thank.

Face to face or online, you helped me through this time, and I want to thank you all for this and will do for all time.

So, all the little simple things will come back very soon, hugs and kisses, smiles and laughs will wipe away this gloom

Dad's on the front line

By Raven Valentino

When something seems so far away it seems harmless, but once it's closer everything changes. I first heard about Coronavirus on the news. It was in China. Most people would just brush it off as nothing. However, it wasn't until it started travelling around the globe that I started to pay attention to it. It began to worry me, and that was when I knew it was serious. I live with someone vulnerable to the disease, I knew that I had to follow all the precautions carefully. I was worried about the consequences of this virus. I began to get anxious enough to ensure I actioned many things.

Where ever I went I took hand gel with me, and if I was at work I would wash my hands after every customer in the usual twenty-second way we were advised. However, it was at work that you could see the virus take hold, as it had almost become like a ghost town. Our precautions were stepped up a whole another level, cleaning all doors, handles, tables, chairs and handrails with warm soapy water followed by antibacterial spray. This was to ensure that we were trying to keep staff as well as the customers safe.

This made me worry more as it had just become serious but I found it also made me more cautious, it made me think about everything I was doing and how I could avoid having contact with both staff and customers. It also made me aware of how little I washed my hands, as I was doing it every five minutes or so. Enough to see the effects of the soap on my skin, dry and crinkling. I had never used so much moisturising hand cream either, in fact, I'm not sure it was something I usually used at all.

Shops were beginning to close down, and people were panic buying I feared that my family may go without. Worry was becoming a significant part of my day now and I wasn't sure how to package up those feelings and deal with them.

When we were walking through the supermarkets it was eerily quiet, footsteps echoed. But finding even the simplest of supplies was almost a mission in itself, and made the panic in me worse.

Watching the news, the data and science facts made my heart race. An unfolding of an apocalyptic fiction movie in front of my eyes, on my TV, in my room. I didn't want to hear about it. Social media might have even been worse. Political debates were occurring there and some of the facts were more fiction. Maybe this is ignorant but I'm sure I'm not the only person who used this strategy; I found I coped better not knowing about it. I would only watch the news if I wanted to update myself on things I had missed.

My Dad is an Orthopaedic Specialist and had to go in to work before lockdown. He was pretty much on the frontline, so I heard everything first-hand about how bad COVID19 was. This made Dad a bit more cautious, and he put some rules in place. We have to get changed into clean clothes as soon as we get in, that's if we have been in a public place or around other people. Mum is still working as she is a carer. She is so brave and has my full admiration. She is potentially exposing her body to the virus. Dads handwashing rule was not to be broken. Every time we came into the house, this was the first thing to do thoroughly. He looked stern as he gave us a lesson on washing our hands. Twelve months ago, that would have sounded like the craziest statement ever.

It could sound a bit obsessive I know, but if that's what

we have to do to protect ourselves from the virus then we're going to have to do it. What has been the most difficult thing about this disease is the lockdown, not being able to see family or even go to work. We were discussing using Skype and the phones just to talk to one another. As it is important at this time to keep in touch and check in on people. Without the human connection to those we care about, I wonder if we might evolve into more solitary people.

Before my work closed down, I did feel a little vulnerable about catching the Coronavirus even if we were taking the appropriate measures. What was I more worried about, being at work or not being at work? It was only at home that I felt safe.

I'm used to staying in, I like staying in. I do wish we could go out if we chose to. Perhaps the lack of freedom of choice is what makes this feel different.

Every Thursday at eight sees everyone unite together as one to offer gratitude, to the frontline staff. It truly is a magical time. Even though we're isolated from one another, we still have the spirit to come together and celebrate the hard work of those who are still braving the virus filled world. I walked through the harbour with Dad and we were emotional to hear people, clapping, shouting and banging saucepans. Fireworks were going off.

The Spectro majestic explosions couldn't stop the virus' spread but our attitudes maybe stood a chance in this war.

Raven is a keen writer. Find her here:

Twitter: @Raven_Valentino

Instagram: @author_raven_valentino

Wattpad: @bluemoonwolfwarrior

Inkitt: Raven Valentino

Ronnie Summers

All gone mad

There's no beans or bog roll,

no pasta eggs or bread,

buy no more than two of each,

well at least that's what they said.

But some have their trolleys high,

bog roll balanced to the sky,

how much more do you need?

is the question why.

It's because we're stupid,

our minds are filled with fear,

what happens if there's not enough,

to wipe my butt hole clear

People stock pile everything,

the Apocalypse is near,

there's nothing left on any shelf

except a Corona case of beer.

Prices rise as people needs,

show a chance for greed,

corner shops robbing the old,

in their time of need.

Sanitizer goes sky high,

a premium has gone on gloves,

people desperate to save themselves

to protect the ones they love.

People crossing paths at a distance,

two meters is the rule,

no shaking hands or hugging,

a nod is very cool.

You can see your family,

through a pane of glass,

babies miss their nanny's,

when will the Virus pass?

Jobs gone, schools are closed, parents left to teach,

but still there are idiots laying on the beach.

So, try to stop going mad,

just do the things you should,

don't do crazy shopping,

stay home and just be good.

Social distancing

Don't come into my space, make sure you leave some room,

at least be at arm's length, when you hold a broom.

You can shout across the road, leave messages on my phone

but please don't come to visit, I'm living on my own.

Yes, we all miss the cuddles, miss a friendly hug,

but you can't come and see me, don't show your ugly mug

The Virus running rampant, but some people are still with

friends,

until they stay on lockdown, the virus will not end.

People dying all alone, scared with no reassuring support,

but idiots take no notice, they have no second thought.

What has to happen, before the idiots clearly see,

must they lose a loved one, or have the country on its knees?

Stay safe, stay indoors, it's not that hard to do,

everyone responsible and that means even you.

The quicker the world stays indoors, the sooner this Virus' war is won,

People; tell your children, your daughters and your sons.

Practice social distancing, nod instead of shaking hands,

cancel all engagements, don't have no party plans

Just find yourself a comfy spot and watch the world go by,

everybody hates it, we could all just sit and cry.

But it's better to not see your family and friends, though it is a lot if strife,

but missing them for a few weeks might just save a life.

One day it will be over, you can share the hugs you've saved,

better to hug them later, than to put flowers on a grave.

Switching off the glitter ball

Chris Quinn

As the news of Covid-19 and the pandemic was declared globally I felt my dance class may need to stop as the deadly virus was spreading rapidly.

Dance is my passion. Endorphins are high when we finish our jive nights. I am close friends with many of the other dances and we meet without prior arrangement regularly, at these social events. The music fills the room through the PA and the glitter ball sheds sparkles onto the floor, chasing our jazz shoes around and losing the race. At the end of the night, I have left behind the tricky parts of life as a ward clerk.

Working for the NHS is something which fills me with intense pride. Making a difference to lives is so rewarding and I love to feel I have an impact. I'm rewarded regularly at work through praise, awards and the occasional cupcake from higher management. I love my job. But yes, it does sometimes need to be left behind, it can be quite the challenge on some days.

Debates were going on in the dance community about continuing to operate or not until the decision was made by the Government. Closed. Immediately.

As the UK went into lockdown I found new solutions, thanks to the internet. Logging on, I chose some comfy clothes to wear and set my camera up in the lounge before clicking play on the music station. 'Line dancing,' I said underneath my breath to myself. I began to feel my heart fluttering, with excitement I thought initially but perhaps little nerves due to broadcasting to the entire world.

Starting with old favourites like the Cha-Cha Slide, the Macarena and all, I began streaming into my Facebook. People responded positively and I thought I could make a difference to the dance community as well as the patients and staff at work.

At work, patients who were well enough to go home were discharged in the usual way and very quickly we prepared for new patients who were being diagnosed with Covid-19. Things were stressful and this was a tricky time. The pressure on the ward and the general feeling in the hospital was like we were preparing to battle.

Unfortunately, a family member who lives with me had a high temperature and we were advised to Self-Isolate for 14 days. This involved staying at home watching Netflix, playing on my PlayStation and Line Dancing. I was one of the lucky ones, able to enjoy some total time off, relax and get ready for the time back at work.

14 days had passed and fortunately, I have not felt unwell so therefore I was able to return to work. I have to follow specific guidelines and policies that the NHS to keep safe. I wonder how things might be when we hit the peak without the spark of the glitterball as a distraction.

No rain, no flowers

Louise Usher

I couldn't sit up or lay down.

The pain was easily a 9. Coughing gave it a 15. crazy high. Worse pain than giving birth to twins.

During the day it seemed to get worse, I'm not sure why, but we were all getting a little concerned.

This was me; good with pain.

Nathan was calmly concerned.

So, we called a breathless family meeting and in between coughing hard, choking and gasping for air, we agreed to phone 999. 111 seemed pointless as the second I said I had chest pains they would send an ambulance anyway.

Not many questions were asked before the operator said they were sending an emergency ambulance even though I said I had an ambulance two weeks ago and they diagnosed pleurisy. Carefully they explained that due to my symptoms the crew would be in protective stuff. Fair enough but this wasn't Covid-19, was it.

Two nice ladies dressed in green with aprons, gloves, masks and even eyewear rocked into my bedroom with all the usual machines. We took the dog next-door to Mums, even though she was asleep. He was a pickle.

ECG stickers were stuck on me pretty instantly and I mentioned that I was sure it wasn't heart-related but rather more something to do with the cough. Sats (oxygen) were typically not bad. They had been very up and down on my own machine. Everything else was textbook and they couldn't hear anything in my lungs. So, they were thinking of dry pneumonia.

Scary word.

Impressive, but scary.

We discussed going into the hospital to get a chest X-ray as this had been going on for four weeks now. So long. Might as well. I need to know if I need treatment. But the NHS is strained. What can we do? The crew seemed to think it was a good idea and left my house to start up the ambulance.

Trying to put my coat on was a killer. The pain was a defo a 9/10. Jasmine helped me with a scarf which had frayed ends which I could fiddle with and keep myself calm. The same theory as a fidget spinner. The paramedics started to get me in a mask before removing their own. Crikey they looked different than I expected now that I could see their whole faces. I was sleepy and a little nauseous. The last massive cough took its toll on me. I felt I was being brave giving the pain a 7/10 at that stage but the paramedic raised her eyebrows and informed me 7/10 is high.

I wanted to sleep in the ambulance. But they phoned through to the hospital who informed them I need to go into isolation when I get there and be tested for Covid-19.

This sounds massive.

Scary.

Surreal.

Entering 'side room 1' I noticed 'Covid-19 testing area' on the door. The paramedic opened that door for me to walk through, so I didn't connect with the touch point.

"Ohh look, pure luxury," she joked, "that's Medway for you." She said light-heartedly. The Medway has been a hero many times for me.

Instantly I was alone and the door closed behind me as I decided to sit on the hard, red, plastic chair in my sleepy pain induced state.

Very quickly, another ECG and set of obs were taken. The results were flashed through the window at someone, maybe a nurse, who was writing the results down from the safety of behind the glass. Soon after, I learned there's a ten-minute window for them to get in and out of the room. I felt for these staff, working so hard and putting their lives on the line. I wanted to give her an extra warm smile, but that didn't work from behind my mask, just like my face recognition on my phone, I said thank you a few extra times instead. Here I was, being tested for Covid-19. When we thought it was pleurisy which needed more treatment. Suddenly this apocalyptic feeling felt more real than ever. Even though I still wanted to sleep, I felt it would be a good thing to get tested and head out of here. After lots of texting and a slight cloak and dagger sympathy post on my Instagram story, the Dr came in to charm and fix me. I answered the questions again about Mexico.

"20th January," was when we came home,

"25th February," was when the cough started. They wondered how I remembered that. It would have been my brother's birthday if he were alive, and I was away somewhere memorable on that day, with this brand-new tickly cough.

"Two weeks ago," was when I was diagnosed with pleurisy and the first ambulance came out. And a couple of days after that I was at the out of hours doctor being told I don't have pleurisy but a swollen throat, given steroids and an inhaler. And here we are now. With my 9/10 pain and a clicky rib.

The Dr informed me my cough sounds wet.

So not Covid-19.

And he was mighty pissed off that I was now sitting in a hot zone where patients have tested positive for the virus. He said the paramedics shouldn't have put me in that risk. He also said the steroids I'd had previously were a baby dose and I need a big dose, plus antibiotics, at a minimum a chest X-ray, a diagnosis and to get the heck out of there and home.

"I can't touch your heart or lungs but I can touch your ribs and the pleura is stuck to the ribs. It's very painful. You need treatment. We have to check what's in your lungs." He was friendly, apologised for rushing and before he left said,

"well, I like to think I can touch some ladies' hearts but I think you know what I mean."

Sitting alone for a good couple of hours, texting my son with thoughts and updates I began to wonder how the heck do I make sure I haven't picked up this worrying virus? I heard a knock and called out 'hello'.

"Can you come to the window?" I heard. So, I did and pulled back the grey vertical blinds. The smiling nurse wrote me a note and showed me through the window. "We haven't forgotten you, we are waiting for an X-ray slot :)" they drew smiley faces on the note. I looked at the nurse's eyes; calming and joyful. Super cute. I nodded back to her and tried to look thankful before letting go of the blinds again and walking to the back of this dark, grey room to the plastic red chair.

I lost track of how much time passed, or even that it was the middle of the night but I felt so tired, I was sliding lower on my seat when another knock on the door came.

"X-ray time my love," the friendly nurse called out and walked fast towards the X-ray area. I struggled to keep up with her. I tried not to touch the doors but kicked out my leg to then waft my way through. This was a week when I'd heard the words 'touch-point' used frequently.

The term needs no introduction.

Outside of X-ray was an old man sleeping in his bed. A huge white gauze dressing was on his chin and I tried not to stare.

My son, Nathan was in the nearby waiting room and he said he would come and sit with me while I get my X-ray done. The ambulance crew told him he couldn't be in the room with me as it was the hot zone. That meant only suspected Covid-19 patients and their medical team could be in there. Poor Nathan had to wait out in the empty waiting room for hours. He was great at taking care of people, a true natural.

The day I drove him to college and began that conversation,

"you can say no, or take time to think about it but how do you feel about being my next of…"

"yes," he said without me finishing. And that was that he was now my next of kin. No other family around is and Jasmine would openly admit her twin brother is much better at dealing with a crisis than her. The deal was done.

We don't need to socially distance ourselves as we live together, so I text him to say I was at the x-ray department waiting room which was right next to the area he was waiting in, behind a yellow painted wall. As he arrived he semi smiled,

"alright?"

He had never seen me with a face mask on before, that must have been strange.

Inside the X-ray room, the staff member was in all the usual PPE stuff and was very sweet with me. I took off my necklace and stood at a metal plate in a position that hurt my ribs. She asked me to hold my breath as she took the image.

"I'll just check that, take a seat," she said and soon after came out and wished me well in an empathetic manner, which worried me. What had she seen?

"I best get back Nathan," I told my son as I crossed the barrier with the words HOT ZONE COVID-19 DO NOT ENTER on it and began to look for side room 1. I couldn't find it. I felt I was a risk to others and I felt myself getting stressed and irritable, which was the last thing I wanted when these lovely staff were trying to help us all. A small, oriental looking man dressed in green scrubs saw me looking lost and flustered and asked if he could help, I explained and he got up slowly from his plastic chair and told me to follow him. He walked just as slowly as he stood up. I didn't want to be putting other people at risk; why wasn't he walking faster?

"That button," he pointed, "that button," he repeated as he nodded to one of those buttons which releases the door.

"I don't want to be pushing the button," I said right after I made a 'tut' noise with my mouth.

On the other side of the door were two male paramedics checking in another patient, I looked at them, they looked at me, I hoped they could guess I was a potential Corona patient by the mask I was wearing, they shrugged and I shrugged too,

"I'm meant to be in isolation, room 1, he's taking me here, this isn't right!" I was a little snappy and disappointed in myself. They pointed to a door and as I went through it I saw my familiar team of Dr and nursing staff and they kindly said,

"in there love," and pointed to side room 1.

"I didn't want to infect anyone!" I said grumpily as I allowed the door to close behind me; back to solitary confinement.

Exhaustion was setting in. Either it was that or the Tramadol was starting to work as I felt sedated. A bed would have been lovely in that room but I felt like a brat with that thought. I fiddled on my phone for another twenty minutes and then a knock on the window alerted me to the Dr, who used sign language to tell me he was going to phone me on my mobile. For some strange reason I held my phone up to the window, pointed to it and mouthed, "this one?" and he nodded and gave a thumbs up.

Taking to my seat, the phone rang with an unknown number and it was obvious who that was at 3 am.

"I don't want to alarm you," he started as my heart sank, "but every time I come in there I cost the NHS in PPE so this is easier. Now, we can see on your X-ray there are changes in your lungs which is a lower respiratory tract infection. So, it's needing two lots of antibiotics and very strong steroids that I can give you. You need to take an inhaler and use it through the day to open up the airways. You don't have to wake up to use it but you must use it regularly as your oxygen is quite low. And then we need plenty of management for the pain. So, I'm going to get your medicines and then you can go home. Is there someone we can call to pick you up?"

"My son is out in the waiting room, he's been there all night, he's a grown-up and he drives,"

"Ok, what's his name? I'll go and talk to him. But listen please, this is very important that you realise you are very unwell with this pneumonia infection and you are a sitting duck for the big virus and you must self-isolate for twelve weeks. Please keep the mask on until you get home and then remove it and throw it away. You must go straight home. Catching anything more will be a worrying time for you," he explained with a wonderful tone, clearly, understandably and with great care. I felt worried. What if I now had picked it up?

I have both pleurisy and pneumonia. No wonder I have been feeling unwell. I was going to take every tablet they told me to take, rest, drink water and get rid of this infection, then I can make great use of the time to write the books and sort my house out ready for the renovations.

I am afraid but, surely, I will be okay.

Saturday morning, I woke at the usual time despite leaving the hospital at 3.30 am. Still tired and feeling rather odd after the events of last night. Suddenly this Coronavirus was a little too close to home for my liking. Rather than being something you hear about on the news from the armchair, this was me, sitting in a room, precautionary measures in place and wondering if I was going to be looking like one of those patients with fishbowls on their heads in Italy, one of the worst affected areas. I'm not quite sure but I think the point of the machines is to level out the pressure in the lungs. The scenes are really disturbing but before last night, my feelings had been similar to when 9/11 happened. You could see it was devastating but it was far away, even though everyone seemed to feel some kind of impact.

But here I was, hoping that my hand washing efforts and mask-wearing had saved me from picking up even one tiny virus from that hospital room.

Medication was never something I like to take. I avoided it mostly during my flare-ups of Crohn's disease and ankylosing spondylitis. I'm a bit of a mess really, mostly autoimmune stuff. Except for the recent half diagnosis. My smear came back abnormal just over a year ago and that was biopsied to find low-grade cell changes. CIN1 they call it. If your immune system is good, you can fight that off yourself without any need for treatment. Even though the gynae consultant said it was a 'failed colposcopy' because my cervix can't be reached (I think it's in a weird place, just like my retroverted womb). They checked me again three weeks ago as I had some abnormal bleeding. I just thought it would be another colposcopy (which was painful enough but I knew what to expect now) but I was wrong. The consultant said bleeding won't be coming from CIN1 on the cervix and performed an internal ultrasound. I had stuff going on in my uterus which shouldn't have been and he needed to inject my cervix twice to anesthetise it before getting inside my womb to biopsy it. I flinched at the first injection and the kind nurse, dressed in lilac, came over to hold my hand, commented on the nice colours running through my hair and I said with a smile,

"This is like when I had my IVF, they talked about all sorts, I should have been working at the FA cup final that day."

"Oh no, I wasn't doing that, I just like the colours," she said, unoffended at my suggestion she was hoping to distract me.

After the procedure, the Doctor explained there is a small risk this might be cancer, but gave me every reason to be hopeful it wasn't. He said he would be in touch within two weeks. My body is trying to fight all sorts it seems. Two weeks was up yesterday. I haven't heard yet. Trying not to panic but wondering if all the clinics will still run ahead as planned. I guess they don't even know just yet. Should I need surgery, I doubt they will do that until my lung function is lots better than this. I must get well. An inpatient at a time when there's a pandemic crisis is not ideal.

Over the weekend the sun was shining beautifully. I felt frustrated not to be able to get out in the garden and cut the grass and do some tidying around. But I simply couldn't. My breathing was tricky and I was exhausted. As well now as being woozy from the medication. At bedtime, I realised my tolerance levels were low. My little dog Harley is a Shih Tzu; they are famous for snoring badly. His noise was driving me nuts. I was in terrible pain and needed to sleep. Gently, I put him on his bed outside my bedroom and as I went to move away he bit my foot angrily. I smacked him and instantly felt terrible. I sobbed, coughed uncontrollably and my children came to help me in my broken state as I was gasping for air.

"It's alright mum, we will take him in with us," Jasmine said.

"What do you need?" Nathan asked and I couldn't reply, I wasn't sure. I think I needed to be out of pain and they encouraged me to take more meds to be more comfortable and try to get some sleep. I became reliant on that medicine over the next few days.

Raising my twins as a single parent was a pleasure, but hard work and I hadn't imagined they would need to take care of me at the age of twenty like this. They seemed comfortable in this role. Pure love and their caring nature coupled with concern for me was what seemed to be motivating them.

So, this was it, I was in quarantine. Fourteen days sounded like a long time and I was pleased I had caught up with my friend Deb earlier in the day before the ambulance situation. We always had a lovely time together and I was going to miss her.

Sunday, March 22nd 2020

Mother's Day in the UK was here. People were nervous about being around other people and some wore masks. We were careful not to go near anyone due to my illness. The children gave me a wonderful day with gifts and heartfelt cards, a cooked breakfast and a drive out to my favourite beach to walk a little while I had some breath. My pocket carried the inhaler they had given me at the hospital. They said I would need it due to the lungs needing 'opening up'. Going out alone became a fear. The coughing fits were worrying and as I felt the head spins that accompanied the lack of oxygen, I hoped I would be well enough to go out alone again one day soon.

The Government announcements were coming at 5 pm every day. I never, ever watch the news. Never. To me, it's too negative and as a sensitive soul, there's not a great deal I can do about some of the state of the world and just feel upset. I'm aware that's a little ignorant of me but as I live my life through the law of attraction, its teachings advise us against such things.

Disaster movies generally have a scene where the mother comes into the living room holding a tea towel drying her hands, to watch the news with the rest of the family. The president of America announcing some drastic steps and measures to combat the crisis. Well, now it seems I might have turned into that character, yes, I'm watching the news.

March 23rd 2020 – Monday

"There's going to be an announcement today, Boris Johnson is going to address the nation," Nathan informed me.

"What? About what?"

"Dunno, some people reckon we might have to go on lockdown as Italy and Spain have,"

And that is exactly what Boris announced. Our Government were allowing us to go out once a day for exercise but otherwise, if you could stay home you should stay home. A certain air of relief was in our house and even though none of us could explain the reason, it was a definite positive. Our dance classes were all cancelled for the foreseeable future and I finally felt I had permission to just lay and get well. Something I had been searching for over the past few years, while I worked in desperation, was the opportunity to stop for a while. Now that time was here. I felt some relief.

March 25th 2020

There has been news that the massive Excel exhibition centre in London is going to be turned into a temporary hospital with 4000 beds and a morgue. This is a very sobering thought. I have been to the Excel for a YouTuber convention, Summer in the City, as well as some other events. The place is huge. A couple of videos have been shared online to show us what's going on. To imagine people laying there, fighting for life-giving oxygen is a very tragic thought. An ice rink in Hull, North England is going to be used as a mortuary too. Something similar was done in Spain and this is now becoming full of body bags with the deceased in them. This is very real. The whole world is affected, the planet is shutting down.

My physiotherapist took my appointment from a physical one into a telephone consultation. I was doing ok really, I knew which exercises I should do to keep the Ankylosing Spondylitis (AS) in check and stop me from turning to stone. He was impressed with my range of motion and seemed to think if I kept up with my motivated active lifestyle, there might be every chance of me continuing to keep well. He would phone again in a couple of weeks and see me back in the clinic after all this was over.

My kitchen designer was due to come out but instead emailed me and used that term 'furloughed' for the first time. I thought he was talking about racecourses until I realised that was a furlong, a measurement. The team at Wickes who were designing my new kitchen were off work for eight weeks, unable to deal with any correspondence. That was a big deal, quite a worry.

Deb phoned me.

"Work is closing today, I'm on my way home. No idea how long for," she told me. My heart felt heavy in my chest and sadness engulfed me like a blanket. Her better half didn't approve of our friendship and we usually only spoke while she was at work. I knew this was going to make things hard and I couldn't begin to explain the difference she makes to my day, every day. She lights me up and my eyes look more alive with her around. I would miss her terribly.

"It'll be ok, we can facetime and that," she explained.

"How will you manage that? Really?"

"There's always a way, you'll see," she said hopefully.

She never did Facetime.

March 26th 2020

My steroids finished. I hadn't been sleeping. I wondered if this was a symptom of the tablets. I felt woozy from the pain meds, which are helping me deal with pleurisy, and possibly with a little bit of the mind chatter which is going on deep inside.

The frost was heavy on the rooves I could see from underneath the gap between my window sill and the blackout blinds. My hairdryer had prevented the blind from closing down fully, allowing me to peep at yet another sunny day. Mrs Usher SNR used to say, "the sun always shines on the righteous." Today, we lay her to rest. My first mother in law. I met her when I was just sixteen years old and had fallen in love for the first time. My twins' father was my first love, first kiss, first everything. History now, long ago in the past.

Due to the coronavirus, the Government has stopped all weddings and other social gatherings like this, but not funerals. Only the immediate family may attend. Which is what we were told by my ex-husband when my son spoke to him. We were not to attend. Very upsetting of course and the twins would have liked to pay their respects. The three of us decided we would go and visit a couple of hours after the burial (if we weren't aware of any new law being passed which meant we couldn't drive out the twenty minutes up the motorway that we needed to drive to get to the churchyard).

Nathan is always very caring around things like this. He always steps up for my hospital appointments. He was keen to go to the supermarket for me (we did need some food) and he wanted to choose some nice flowers, if he could find any, for his Nan.

As he parked his car, he text, 'it's a ghost town,' and fifteen minutes later another text came, 'this is a nightmare, queue right round to Costa,' Jasmine laughed as it was to our group chat and she remarked how she just read his text highlighting the ghost town. Nathan tried to help an obese lady in an electric wheelchair to reach an item from the top shelf which she couldn't reach. We have social distancing in place, two meters or six feet apart from strangers. She didn't attempt to move her chair so Nathan could keep his distance and get her item for her. Thankfully, he is abiding all the rules about hand washing for twenty seconds and using hand gel. We all are, to the point we have very dry and sore hands and need to use diprobase to try and keep the skin moisturised.

Everyone is doing what they can to help stop the spread of the virus, or 'flatten the curve' as it's been called. This is about the scientific graphs you would see as the numbers rise and fall again. The flatter the curve, the more the people have been unaffected. I'm not sure if they record the numbers of those infected or those who have died. But lots of people are dying right now and it is going to get worse.

Boris Johnson called for volunteers to help the NHS with jobs like delivering prescriptions. He hoped to recruit 250,000 but within no time it was reaching around 405,000. Crazy. My friend who works in a bank (the banks are still open) has volunteered. Brave. She suggested I sign up too. I reminded her I am in isolation for 12 weeks.

Nathan returned from the shop with food and Nan's favourite flowers ready for the visit to the churchyard later. I was sure our journey driving there would be quiet but I was wrong. The roads were not much more empty than usual, people were walking around the streets and generally, things looked much more normal than I had imagined from behind the safety of my windows.

Following the map, we pulled into a local pub car park, knowing the pub was closed and thinking that the owners really wouldn't mind us parking in there for ten minutes while we went and found Nanny's grave and laid some flowers.

Chalkboards in the car park said 'open for takeaway 12-6' and part of me was impressed with their enterprise and part was shocked that there will be contact with other people during this time. It's hard for any business at this time; from the local pub to the major airlines.

We were astounded to see people standing at the back entrance to the pub smoking, chatting to the staff there and being handed take away boxes of chips. But that was their business, or was it? Little acts like that affected the whole world and now things were beginning to become very serious and we all knew that no one should be in contact with others.

Trying to climb a slight incline on the way to the church, I felt myself getting breathless. My lungs had always been amazing. Swimming underwater is one of my absolute most favourite things ever. It's so peaceful under there and no one can disturb you. Suddenly I felt old. Breathless, me?

A square church was in front of us with a beautiful backdrop of blue sky and sunshine. That phrase she always said, the sun shines on the righteous, she was right. She was a good lady with a kind heart and she deserved a funeral with all of her family there. This was very sad. She gave birth to seven children, had fifteen grandchildren and a whole load of great-grandchildren. A typical Nan with knitting needles and strong cups of tea always on the go. She and I used to enjoy lovely chats when I was off work on maternity leave with the twins.

Unsure how we would find the grave, I assumed it would be quite obvious. Jasmine was shocked to see we were not in a graveyard but rather a churchyard. It felt right somehow, she was deserving of a nice place to rest.

We tiptoed through the spring flowers surrounding us, and the fallen headstones in this old churchyard and something began to feel familiar about the place.

"I have a feeling this is where your Godfather got married," I said. There was a particular slope of grass which jogged my memory,

"I've got a feeling I can see you two rolling around there at about the age of two," and the twins laughed. They were never the type of children to roll around on the floor, they were right to think I was making it up. Their reaction reminded me again that it was one of the male family members tickling them which took them to the floor in giggles, they were the type to scream out, 'you're making me do this and it's kinda funny but also torture, and I really shouldn't be rolling around on the floor in these clothes,' but all of that felt very long to explain, as well as seeming disrespectful as we were looking for Nanny.

All of the headstones were old and moss-covered with occasional lettering missing. We looked left to right and could see nothing which looked like a new grave. To our right, towards the slowly setting sun was a flattened patch of grass. My eyes followed it further into the distance and I could see a newer patch of the cemetery,

"look, this looks newer," I said to the twins and led the way, holding Jasmine's arm as I was unsteady from the medication.

We began walking carefully between the graves and at the opening between the two pieces of the fence was a plaque saying 'this extension was created...' and so we felt this was likely the right place.

We saw two people standing at a grave slightly up the hill from us and I whispered to the twins that it was the family and I gestured to a bench underneath a tree for us to go and sit on and wait, as if I was a mother to toddlers. None of us were dressed to impress but that didn't matter, we were here to pay respect. It has been years since I have seen that side of the family and it took them a while to recognise me. From our two-meter social distancing, we exchanged nice words about Nanny before we went and laid our flowers, shed some tears and spoke a few words to her.

"After all this madness is over," seems to be a phrase adopted by many currently, and today was no exception as we were told there would be a memorial service after all this madness is over.

The twins did themselves proud on this day. Unlike me, who felt angry at it all. She deserved better and to be honest, I felt we did too when we heard there were other family members there who were not immediate family. Channelling that anger, alongside the pleural pain, I decided to get in the kitchen, play Kisstory up loud and batch cook shepherd's pie and carrot cakes.

Exhausted, I had to take to my bed for a few more days as my illness seemed to get a grip of me. Was I doing too much? Had I created too much movement? The pain was horrendous and I resorted to taking lots of Tramadol to try and ease the sensation which was making me grumpy. Maybe it was the tablets, maybe something else but my mind seemed to start to harbour unhealthy thoughts.

I'd heard many people were feeling the same, experiencing highs and lows in their mood, and the NHS pop-ups were now including 'how to look after your mental health' on the laptops and mobile phones throughout the UK.

I'm aware of my mental health enough to keep safe. But here I was feeling angry, in pain, unwell, not sleeping, upset stomach, missing my life a little, missing exercise, dance, my special people. Some friends grew closer and messaged lots, while others seemed to have vanished into the ether. Understanding what others might be going through wasn't so easy either, there was time to message, so why didn't they? But we all had our own, very real, challenges at this time. People worried about money, work, the economy, buying kitchen rolls and eggs. Even the simple stuff felt like a punishment, missing their routines and workouts, setting the alarm and singing along to the radio on the way to work. I guess we all internalise that differently and I do try to be one of those humans who is understanding of others, whatever their journey is. But sometimes my journey feels harder than I would like it to. Sometimes I wonder if I have the strength to carry on, knowing my health is potentially going to make things difficult going forward in later life. I pull out the drawer next to the bed and look at the next dose of Tramadol I should be taking and close the drawer again quickly.

Should I move these? It's a big box.

No, I can do this. I'm going to sleep, face a new day with heat treatment and paracetamol and kick this poxy pain into touch so hard until I'm smiling and positive again. I was determined to sort out all aspects of my health and do whatever it took to get there.

I even put my pen down for five days. Involuntarily.

I didn't journal my news from the hospital which arrived the day after the funeral. And the smile on my face wasn't as wide as it should have been, but that was external. Internally I was feeling incredible. And shocked. The endometrium was clear. No nasty cells were detected. Frigging fantastic news; now for the cervical assessment.

March 31st 2020

I know this sounds very negative, and I do pride myself on being positive, I couldn't help but expect the worst.

Maybe it was because I lost my brother at the age of 31, maybe it was because Dad died so suddenly but when they tell you there's a chance it could be cancer it's hard not to think the worst. I had told my loved ones but kept everything else private. So much of my life is online but this felt like one of the things that were just too awful to put out there.

Also knowing how the law of attraction was working made me feel the less energy we gave it the easier it would be. So, nobody knew about the pain had gone through when they injected my cervix or taken a biopsy from my endometrium. I was alone in the chair that day, just before we went on lockdown. I was confident I would get the best care from the NHS. Everyone was cheerleading the NHS right now, so different from the grumbling people did before this crisis.

The gorgeous chubby-faced nurse dressed in my favourite colour offered her hand for me to hold after the first time I almost leaped off my bed.

Yes, it hurt. But it was necessary, and I was hopeful we might be able to get some kind of treatment and get rid of it. I told my nearest and dearest of course who were incredibly positive and encouraging, telling me it will be okay and we will fight whatever we need to fight together. That meant everything.

With our incredible NHS under so much strain with the Coronavirus outbreak, I couldn't help but selfishly worry what might happen to cancer patients at this time. I wasn't sure if all of our frontline healthcare workers would need to pay attention to everybody with Coronavirus. But the scientist and healthcare provider in me knew they would still have to attend to other urgent cases.

I had a colposcopy booked. I needed to know if this was going to still go ahead and so I picked up my mobile and called the secretary. Considering they were surely under plenty of stress, I was so impressed with the way that she answered the phone sounded so animated and positive. After explaining my situation, I asked if everything was still going ahead,

"yes, you are still booked to come into the clinic, are you able to do next Tuesday instead?"

"Is there any reason? I mean, I can, but is there any reason?"

"What do you mean?" The secretary asked me.

"Well, I had a biopsy done. I'm still waiting for the appointment to tell me the results of that," I said with shortness of breath. She asked kindly if it was okay to put me on hold while she went and checked. Of course, I said yes, and mouthed silently to my son that she was going to check. He whispered back to me,

"ask if I can come? "

After a short while on hold, with some music playing and intermittently having a coronavirus 'what to do' talk, she returned.

"I have your results here. And they've come back completely normal," she informed me.

While my face didn't show any expression at all, I mouthed over to my son,

"normal," with a frown as if to say was it normal? But my actual response to her was,

"oh, that's amazing, I know I might need some treatment

but that's still amazing news."

We continued to talk about my next appointment which was in a couple of days.

"All these kinds of appointments have to still go on," she informed me. I asked if anyone can come with me and she said no. She said they are trying to limit the number of people coming into the hospital and therefore that means no visitors or accompanying guests.

My son was going to drive me there and back and wait outside in the car park. I was nervous but the appointment was very necessary and super important to my health. I couldn't help feeling nervous about entering the hospital knowing that there was very likely to be somebody somewhere with the coronavirus. I'm not sure if I would've felt the same if I hadn't had this pneumonia diagnosis, but I seemed to be on edge. I just found it very hard to believe I was going to live into old age. This changed my thoughts.

The night before, drunk on prescription medication, I voice noted one-minute message after one-minute message on Facebook to my friend Pip. She had supported me for years now. A diamond friend who is one of those gorgeous humans who is always there for you. She has remained in my heart since 1987 when we met.

Reaching out to her that night brought us to a conversation that I'm sure we would never have had pre-Covid-19. I began to wonder how many other thoughts and feelings might have been addressed in these times which would otherwise be suppressed or ignored. How hard must it be to be stuck in a house where you aren't happy? How amazing must it be to be locked down with someone you've craved to share time with but never could because life took all of your time together. Time had never been more selfish for some.

Pip had my total trust and I knew she loved me dearly. She could say anything to me and I never could have taken it negatively. Possibly she is the only person in the world who could have told me that I shouldn't be afraid to open up my heart again. I'd never considered that I closed it as such, but after replying with voice notes, I realised she was right.

"Why do you always keep people at bay? "Pip asked me,

"do I? Do I do that? I just don't think I've met the right person," Pip very gently reminded me how young I was when I got married, how he treated me, why I might be reluctant to let anybody in. She also said, "your kids are growing up now, what about someone for you?" For some reason, it triggered something inside me. I've never thought that I was deliberately closed. But Pip evoked emotions with that conversation. I think without the pandemic hanging over us we filtered out words more and held back on things that we would otherwise have thought or said. Pip had made me think. Did we talk like this because we had time to sit up and have a late-night chat on Facebook now? Whatever the reason, we did have that conversation and she did begin to plant the seed in my head. I didn't feel like I knew the answer, but even asking the question was something brand-new for me after so long of believing it would be me, myself and I. This pandemic has a lot to answer for in response to one's thoughts.

"The kids won't be around forever honey; don't you want someone for you? Someone you can share your life with?" she asked and my initial reaction was, 'nope' but as she went further and deeper, she got me thinking. What if this someone was amazing, what if he made me happier rather than more stressed and uptight? I'd been so afraid of getting those barriers smashed for so many years now which was possibly the one reason I'd stayed in the single world.

Could I consider opening up my thinking of letting

someone through my force field of doubt? Only Pip could say that to me. Only the Corona Virus could stimulate this thinking pattern.

I felt open. For the first time in years I was thinking; maybe.

Pip and I were talking about health and the future. I told her that I had partly good news and I was just waiting for the rest of the results before I could look forwards. It was a very strange time to be having this conversation because it felt way too much like normal life or playing down the pandemic. But usual life, usual tests and health problems were still carrying on. And I couldn't help but to be afraid of my mortality, which is something I thought I'd lived with now for several years. My chronic diseases were so much better now that I knew how to live an active healthy lifestyle with good nutrition in it. Since applying these techniques, I felt so much better mentally and physically, sleeping better, all the things that I believed might make me better.

"Ready Mum?" Nathan asked as he swirled his car keys around his index finger.

Getting into the car felt strange as if we hadn't been driving for a year. There is no way my arms would've moved enough to drive myself, due to the pain from pleurisy. So, as I fixed my sunglasses to dim the brightness of the gorgeous day, I was thankful to my son for driving me.

We spoke in great length about the best ways for me to keep myself safe throughout this appointment. Including not sitting inside the hospital any longer than necessary. We pulled outside the drop off zone and waited. I wound my window down slightly as it was a warm day, the sun's rays were streaming into the car. We heard an announcement come over the tannoy right near the entrance doors which normally opened automatically, but this day they were covered in signs saying 'entrance closed please use the main entrance'. I didn't want to use the main entrance because you had to go through a red and white striped cordoned off area and follow the path in, anti-bac your hands and pass many people. I was afraid. I had pneumonia. I did not want to catch Covid-19 in the hospital. Listening to the tannoy, we heard it say "please do not enter if you feel you have any symptoms of Covid-19, please go home and dial 111," and we looked at one another and pulled a shocked looking face. Not long after that, I saw some people going in and out of the door and I figured I could do that as well.

I had no intention to touch anything. No door handles, only the anti-bac on the walls (which turned out to be empty, thankfully I had my own in my handbag), I wasn't going to be within two meters of anybody, except when I was having my tests. My stomach was in knots. But this test was of paramount importance. It all felt very surreal when the rest of the world was fighting for their life against a virus and here I was trying to make sure my cervix and I were in tiptop condition. Or at least a condition that would allow me to survive a longer life.

As I entered the outpatient department, with my letter in my hand, and tried not to cough, I noticed there was a lot of cordoned off areas. The receptionist looked up at me and smiled with a kind face.

"On the cross please," she pointed to a cross, made out of tape, stuck to the floor which looked approximately two meters away from her. Of course, I obliged and stood on the cross looked up at her and smiled, as you would do with anybody working in this scenario. The NHS workers deserve kindness. I had to yell my name, date of birth, GP surgery and telephone number across at 2m distance. She invited me to take a seat, I made sure I wasn't sitting near the other patient and her partner who had gone in with her. I text my son to tell him there was another patient with somebody in with them.

'Are you kidding? That is well out of order. Do you want me to come in?' He asked me and I replied,

'no, I will be fine'.

The conversation between me and the medical team saw me thinking they were frustrated that patients were not coming into hospital as much as they should have been due to fear of catching the virus.

"Young girls, with big treatments, they aren't coming back for their follow up, too scared,' they explained to me.

As soon as my biopsy was done, I was out of there. I could have done with support from Deb on this day. But not even a text arrived. I kept strong, but this did make me sad. I missed her. We were all missing our people. Doubtless she was missing me and the laughs we had.

April 1st 2020 – a brand new month

Spring was happening outside.

Daffodil trumpets looked orange and the sky somehow

seemed bluer than before. Social media, mostly Facebook, had posts from people speculating about the pollution improving and photos of space looking back to the world as clear as day. Comments about the ocean improving and the reef over in Mexico, where we luckily spent two sun-filled weeks in January just before all of this started, having some time to repair, is a hypothesis which I love. This is something I would be so happy to see. Is the world going have a chance to heal a little while we all sit looking out of our windows instead of driving our cars, littering with plastic, cooking more from home and using a whole less 'throw-away' items?

I did even wonder if there might be some way this virus is complete poppycock and instead 'they' (whoever they are) are getting us to save the planet. A conspiracy perhaps. Imagine that, "you all must stay inside for a few months to save the planet," versus, "stay inside else you might die," well, I know which one would motivate the masses more. But I am confident that my fleeting thought of any conspiracy is total fiction.

Heal, planet, heal. While we are going through this time.

A new normal, it's being called now. Yes, this is no longer abnormal. This is our lives. We work at home, we clean, empty cupboards. I frown at the copious amounts of comments on social media detailing how bored people are. Bored? You are bored? I'm never bored.

Always busy; never bored.

I was due to have new doors fitted in my house before lockdown. We pushed the date back more than once and my chippy touched base with me checking if I will be finished with quarantine in ten weeks' time. I text back to say yes and he sent a photo of his wall planner with 'Lou DOORS' on it for a date in June. This seemed an age away. June? I was excited to get my house renovations done – such a pity everything was on hold. We all work so hard to pay for these little boxes we spend time away from, it is nice to be home.

Even though I was busy with work, there was an opportunity for conversations with the children that I might not have otherwise had. Jasmine kneeled on the edge of the cream leather sofa and flicked open the blind with her index finger. I was willing the blind not to break, they were so fragile,

"oh Mum, bins?" Jasmine announced.

"Ahh you mean do-dammer?" I said and we laughed in unison. When the children were tiny, before we lived in this house, they confused me one day. They both came into the kitchen pointing to the front room saying, 'do-dammer' over and over. I had no clue what they were talking about so followed them and they showed me the dustbin men were outside. From that day to this, the dustcart and its activities are called, 'do-dammer' and even 'do-dammer day.' Other families have unique names for Grandfathers, each other, something. Always with a great story behind it that will make them smile.

The bin men wore PPE masks, gloves, aprons. Very brave.

Walking Harley that evening, we ended up talking of stories of 'cutting their teeth', what it means, who cut first and me missing that because I was at work (I would have lost a bet that Nathan was first).

On the walk we admired the gorgeous house of the local McLaren owner, took photos of chalk rainbows drawn on the pavement by children and laughed about things we might not have made time for before.

Lockdown saw conversations exchanged between families that otherwise wouldn't have been shared. Today was a good day.

Monday April 6th 2020

People report that they are getting some sort of routine. Some people are even now enjoying their time at home and I feel no different to that. I have tried to create some home workouts now and aim to get some strength back again. The past couple of days walking has shown me how weak I have grown over the past month or so. But my body showing all the signs of not wanting anything but rest. The pain was still bad but I was going to have to push through to improve from here.

It's Monday, a day for my focus time online, always, and today was no different. I tidied the office a little and settled down to write at my desk with the sun streaming through the white slatted blinds. They needed dusting. But usually, and now, there's no such time for these things. And when there is, I'd far rather be out making memories than dusting.

All-day writing. Aside from waking the dog on the one allowed walk of the day, oh, and dropping a little bag over to mums' house, which I hoped would excite her. A banana, a mini pot of marmalade and a tin of pilchards in tomato sauce which no one wants in my house. I likely bought them as a low carb snack at some time a year ago. She had a carer into her house half an hour before and I had explained to her on the phone that she needed to stop brushing her permed hair. It just looked like white candy floss,

"you don't brush perms mum, else you end up with a load of frizz, just wet it down and leave it curly, it'll look nice, you will like it." I encouraged her not to look like Worzel anymore and as she answered the door, there she was with naturally drying, curly hair. I was impressed. She listened to me! For a day anyway. And she was clean. Without the carers, she doesn't wash and me going in there is not an option right now as she has had a fair amount more contact than me. That's potential exposure to the virus.

She will tell everyone,

"she can't come I here and give me what she's got," but the reality is, Mum, you are way more likely to have Covid than me and I can't afford to catch it any more than you can. Those words echo, "Sitting Duck," the doctor in the emergency department told me.

Being a carer in these times isn't easy. So many families torn between duty and survival. We were fortunate, with the professionals on hand, and my Olympic sprint to leave the front door before Mum opened it, we could manage. If only Mum tried to understand.

April 7th

So many people online had been describing this period as a rollercoaster. It was very much full of ups and downs. Plenty of worldwide white knuckles for sure.

The fact that I hadn't found anybody who wasn't saying the same made it very real. This was strange. I was no different from anybody else. I was generally pretty good at knowing how to pull myself out of a down mood, and even share that kind of stuff on my YouTube channel every Monday. For some reason no matter how amazing this time was feeling on some days, it was not feeling good this day.

No real reason, it's just how I was feeling. And I decided not to try and figure it out, but instead, I tapped into talking. My daughter is a good listener, and I fell into her arms and sobbed.

I had got to the point where I wasn't sure that I was likely to get any better even with my best intentions, so had to take an early night and make sure that I got some sleep.

Thankfully for me, I had a fantastic conversation with one of my friends just before I went to sleep which changed my mood, and I managed to sleep for a total of seven solid hours which made me feel so much better.

April eighth, a Wednesday.

Opening my diary, I realised that this was the day before Maundy Thursday, which was a strange time to think it's going to be Easter. One day was beginning to feel like the next. I didn't even know if we were able to buy Easter eggs at the supermarket. Besides which even if we could I wasn't going to the supermarket. What was I going to do? Yesterday was the highest death rate we've had so far. And our Prime Minister was in intensive care. He had contracted Covid-19. I had visions of him being intubated and in a coma. But he was breathing by himself. I was pleased to hear that because he's a face that I have been watching on the news, and I felt as though I knew him a little. But then the news filtered through via Facebook that my friend I've known since I was five years old had contracted the virus. She lived in the next town to us and worked in a local hospital as a ward clerk. She was at home and she was breathing by herself. I couldn't help but worry about her as she is so lovely. I never forgot the time that she went in to visit Dad when he was in ICU the day before he died.

Everyone knew someone who had Covid-19. Just a few weeks ago people were joking about Corona with a bottle of Corona beer in their hand and it seemed a laughing matter for some, not the next-door neighbour tragedy it was now.

Easter Day

The strangest Easter was upon us. The churches were closed, the tulips bloomed toward the sun just as usual outside in the graveyards. We walked Harley the shihtzu up to the door and read the notice detailing the church being closed. I'm not quite as committed to the church as many but I do like to go at Easter and Christmas and celebrate and give thanks. Last Easter we were in Wales, staying at a beautiful barn conversion. The old stones which made up the wall outside were typically welsh looking and the view was lush and ever-changing, watching the shadows creating an original view at each hour of the day. I loved this place. We stayed there several times now. I loved to travel. When might we be able to travel again?

That Easter in Wales, I took mum to the local church in the nearby village. A fan of Wales and the Welsh, I wasn't sure why I'd seemed to forget just how famous the Welsh are for singing. The choir were amazing and I felt in awe of their six-part harmonies considering how few of them there were.

The choir I sing in as an alto (frequently an accidental soprano) is huge in comparison. When all of us are present there are almost eighty of us. We are good. This is a university choir which I joined back in year two of my science degree, to help with the stress. I'd say it worked. But I'm not sure I would say that after the very first rehearsal.

I was given sheet music. It was Mozart. Mass in C minor. The first movement was Kyrie. I'd not read music since I was at school, and that was just a one-line melody, never a six-part harmony. So, of course, I was quickly lost, close to tears, frustrated. We took a break for tea and coffee, biscuits and cake. I drank water. I felt concerned about my Crohns misbehaving while coupled up like a bully with the frustration of trying to follow Kyrie. During the break, our choirmaster came and introduced himself,

'We know each other from email, don't we? Louise?' he

asked.

'Yes, hello Nick,' I shook his hand and I was visibly impressed that he spotted me in the crowd, recognized me and made the effort to say hello while he had a queue of choral exhibitioners waiting to speak to him. I'd hoped he hadn't seen my poor efforts of trying to move my mouth in the right places while taking care not to let anything audible leave my lips. I'd hoped he hadn't seen me biting back the tears of frustration. Getting to know him over the next few weeks, I firmly suspect it is very likely he did notice. After I smashed that first concert (I called in a lot of favours from some amazing musical friends), Nick hunted me down afterwards at the post-concert drinks.

'Every time I looked over to bring the altos in, I looked at you and you were like,' he gestured with his hands and opened his mouth to demonstrate me singing with confidence and gusto.

'Well, I worked hard on it.' I said,

'but Louise, this is Mozart!' his face tightened to show a dazzled expression.

Nick and the choir changed my life. He will never have any idea how those early words of encouragement made me feel. The me who was afraid of not being good enough in life became the me who was going to overcome obstacles in the next few years to chase dreams and stand up and be counted without crumbling, no matter how tough things got.

Easter in Wales was lovely. The choir saw Mum expressionless as usual.

I could never step into Dad's shoes. He was the most patient and giving husband, making a rod for his own back by doing everything for Mum. The cooking, cleaning, driving, shopping. I and the twins have memories of him telling Mum to go sit watch tv, after the dinner he cooked, and would stand at the sink singing his heart out to the 1960s songs on Smooth radio while doing the washing up. We chose to stay in the kitchen diner with Dad while mum watched TV.

Of course, a rod for his back now meant a rod for mine.

Since Mum was taken into hospital with a NSTEMI heart attack, she hasn't been home. We tried our best to take care of her from our house next-door. Yet since my pneumonia and being instructed by the doctor to be on total quarantine, Mum was still going out to her day group which I was flabbergasted to know remained open for a couple of weeks before lockdown. I felt it should have closed but the decision-makers were between a rock and a hard place, I understand that. The elderly people needed stimulation and the family carers likely needed to continue working. But it was potential exposure to the virus, which frightened me. Usually, I would go in to Mum twice a day and take her obs but there was no way I would do that unless it was necessary at this time of lockdown. She was able to take complete care of herself. She could cook, clean, dress and use things like the TV without any problem. Sometimes she decided not to wash. She told me it was because she didn't feel like it. I explained how important hygiene was at these times but she seemed to stop listening then.

Carers came in three times a week to bath her and put out fresh clothes for her. She wouldn't get in the bath with my encouragement.

"Can't be bothered," she would say. So, instead, I brought in the professionals who ensured she could be bothered.

Trying to explain to Mum that I would be phoning her,

talking to her from the safety of the front gate (over four meters away from her front door), placing food on the doorstep and asking her to try cleaning her own house instead of me coming to do it, was a little ignored. She told all of her friends this was about her not catching it,

'well that'll probably be my lot if I get it and I'm not ready for any of that yet thank-you-very-much,' my 80-year-old mother told me. It was a miracle she survived that massive heart attack. Her troponin levels were at almost 7000 which gave the health professionals wide eyes. An ex-smoker of over 60 a day, Mum was lucky. Now, with heart failure beginning to show signs, I couldn't help but think it was a miracle that she was as well as she was. True, she is vulnerable, but so am I. Mum always had the talent to encourage situations to be about her. It was irrelevant that I had pneumonia and was categorically told by the emergency doctor that I must be on total quarantine from this highly contagious virus. Mum thought nothing of exposing herself to people unnecessarily, talking to neighbours over the fence without distancing. And every time I went to deliver food to her door and knocked it, I had to run up the path as if it was a 100m Olympic event.

'Nah, she can't come in now, she can't go giving me

anything,' Mum would tell her carers.

I would phone twice a day and encourage her to speak to her friends on the phone, which she enjoyed,

'spoken to Dawn today mum?'

'no, she didn't answer,'

'oh, maybe she's up the garden,'

'maybe her daughter has taken her out,' was that a dig at me not taking her out? I was confused and beyond hurt now, 'where to Mum? The pub? The pictures? No one can go anywhere Mum, everything is shut. We aren't allowed to go anywhere except for the food shops or medical stuff,'

Monday 13th April came and it was about that

time of night when we could expect to hear from the Prime minister.

He was somewhere recovering from the virus, so another person at number ten gave the press conference. They were always so polite to the press. Despite sometimes listening to the questions which were designed to irritate and antagonise. Always polite.

By the time the conference finished, we were left confused. What was that news? Were we still in lockdown? The guidance was a little diluted I felt, not very clear at all. Even though everyone felt they knew there was zero chance of lockdown ending and we would continue with the new normal for now.

Would life ever feel the same again? When would we be back out in the Nando's queue, complaining about the length of time we had to wait? When could we casually ask the rest of the family, 'shall we pop to the pub for lunch?'

'Fancy a Starbucks?'

My local countryside walks fill me with gratitude just now. This evening the sunset was breath-taking as it lit the distant sky with orange tones and created silhouettes like lace with the trees on the horizon up the hill. Evening bird song from the blackbird on the nearby roof was beautiful and I pulled out my phone to record him, knowing that the zoom would make him too pixelated.

I do hope they allow us to continue our evening walks. Today it saved my sanity. My son encouraged me to go. He knew I was really tired from overworking and trying to keep up with both work projects and house things. My daughter needed to use my office and so I took the laptop and headed to sit on my bed to carry on with my work. Instantly I fell asleep,

'Mum, you need to wake up and do your work,' he gently told me, knowing I wanted a productive day.

'when did you want to go for a walk? do you still want to?' he asked.

'No, no I don't but I should as I know it will make me feel better, then I can work later,' I was beginning to think I was working too much, yet I didn't know the answer. Being self-employed was difficult. Money was tighter than usual. I was always resourceful and I could create income in different ways but I needed to put in the work. Part of me envied those who were still getting paid yet could put their feet on the pouffe and watch as much Netflix as they desired. Everyone thinks they have so much time right now but I have these projects to fulfil as well as trying to empty my house of accumulated junk, burn old paperwork and sort the stuff that wouldn't even sell on eBay. Hopefully, our bundling work would begin 'once this was all over'.

April 15th 2020

Where was my head at? I wasn't sure what I was feeling.

I woke okay.

Determined to get straight to my desk and figure out some forward plan into this career I was feeling focused on during this time. It was a creative's dream, this quarantine. We had a chance to catch up, write the book, create the online business, read more, paint, clear out the junk, cut the garden down to within an inch of its life and throw away pasta bake sachets dated 2019.

Was life so different for me right now? Working from home as usual, just with the children here all day with me as a distraction, not going to Starbucks with my laptop, journal and pen in tow. Not seeing my usual people converse with, to hug, to feel, was difficult and today was no exception.

Trying to tap into my muse, I stumbled into Twitter land on my iPhone. So many people had muted Twitter. The political debates and hatred were quite damaging to many people, including us creatives. Someone who has been an online cheerleader of mine for years now replied to the creative Tweet I wrote as a visual piece of writing which came to me as my brain headed into subconscious slumber the night before.

"She heard those unsaid words loud and clear," I wrote on a piece of lined paper in lilac gel ink and photographed it ready to make a Pin for Pinterest. They loved to re-pin little phrases and sayings like that over there and I saw no harm in sharing it to Twitter. In pure fun, my online friend tweeted back,

'you sure that wasn't the voices in your head?' and I grinned as I read it. Clicking to reply, I noticed his account had a tweet of a political nature in the feed next to the message he had sent me. Mentioning that an MP was laughing about all of the deaths, I felt astonished as, although I'm no expert on politics in any way at all, I had noticed the empathetic nature in which all of the addresses I have seen from number ten downing street had been conducted. I liked this. Such a tragic time of worldwide sadness. It's beginning to get to the stage now where the virus is something closer to home now. We all know someone who has it or knows someone who has a relative who has died. We all seem connected now, like a spider diagram of sorts.

I felt compelled to click on the Tweet and watch. Piers Morgan was interviewing the Care minister and claimed she was laughing. Piers was doing the typical newsworthy thing of trying to close down a yes or no answer in a very strong manner and Helen Whatley wasn't laughing about death at all. It makes great watching for those who like drama on the TV I guess but then it inspired Tweets like,

'Yeah that's the tory way, laughing about those who have died,' and frankly, it made me angry.

Why can't we try and be kind in times like these? No, sod that, why can't we just be kind? We should be pulling together, in true disaster movie standards and with a good bit of British 'let's pop the kettle on and have a cuppa' mixed in.

Ok, yes, I have been watching the news and usually, I don't. I'm trying to get facts to ensure me and my family are playing by the rules and figuring out what is going on in the world so I can keep them safe and cared for. Feeding five mouths, six including the dog and juggling paying for two homes while earning less than before is not an easy task. Then, silly me, I signed up for an online course to grow my YouTube channel as well as to write this book (not a complaint; fantastic therapy) and have over-committed myself. Reality is I should probably be selling all of our unwanted goods on eBay to bring in a little extra money while also clearing the house as much as I can before our house renovations begin.

Kindness should be the priority of these times. My friends and I are having moments when we are struggling to remain normal with each other. I've taken to using voice notes for all the enquiries I'm having coming in regarding the book. I can cover so much more in much less time and it saves my eyes for more important screen time.

Things are not normal, this is the new normal. We are finding it's taking creativity to keep us close, which is partly quite exciting. We are making plans to go out and we talk about things we want to do 'after this is over' but can't currently do. Julie and I have decided so many things. A trip to London west end, a spa weekend, maybe a week away in Spain too. Other friends and I share pictures of where we are and what we are doing, the occasional phone call and some voice notes. Videos have been very handy too, especially for stimulating my creative brain, being a visual person. Above all else, we have tried to be understanding of each other's moods and the myriad of emotions we are all feeling daily as things are changing within us.

None of us can understand why we feel the way we do. Moreover, we are beginning to accept that we just feel different. People online, myself included are sharing the idea that it's okay not to quite understand how we are feeling right now. It is okay not to be okay. It's okay to be feeling fantastic one day and not the next.

Some days have felt wonderful as we wake up to no alarms and most days have seen the sun on the other side of the curtains. We can spend more time with our families, more time exploring the outside of the house on our once-a-day allowed exercise. Taking stock of the little house that we work so hard to pay for yet seldom spend time in. Having moments with ourselves to reflect on where we are heading in life, wondering if we can take that desired path, is a time-sensitive luxury, which this horrific virus has blessed us with. The tragedy of the loss of human lives, of loved ones, is cruel and uninvited. Fear has consumed many of us as we worry about the simple act of taking a breath. Is that person on the other side of the street contagious? Keep away, please.

Unity consumed the world as we mentally held hands in valour against the virus' spread. As we did so, we watched the rivers becoming clearer, the visuals transmitted from space brighter, the pollution lifting, some even said the ozone might repair itself. Animals learnt how to be more natural and we found out how to feel again. Having time to stop and breathe, eat the right foods, sleep as our body commanded and fight for survival was what we are designed for. No reliance on caffeine, sugary snacks, music up loud in the car to pump us for the day ordinarily needed for many people who otherwise couldn't naturally get through the day.

Time. Time to think. To get clear.

Physically, we have to survive this time. Mentally, if we push and keep pushing ourselves through this into new heights, smashing boundaries to shards who knows what we might achieve? Perhaps, just maybe, we can allow ourselves the luxury of thinking things might even be better after lockdown.

When we are given back the gift of our lives once again we will be reunited with loved ones. The warmth of an embrace could grace us with the blessing of realising that while on lockdown we have reset ourselves to open up to feeling deeper than ever before.

No rain, no flowers.

Afterword

How writing made us feel

Louise

As a scientist and a life writer, putting pen to paper was a must for me. Especially as I started to write in the ER while I was being diagnosed with pneumonia. Then I started to ask others to write about their experiences and I wondered if this could be a fab history piece for those in future years.

Writing is my absolute passion and without journaling every day, my mind feels like it wants to explode. It's therapy.

Others have loved my encouragement in writing from life and to me, that is so incredible. Being told you have inspired someone to write is an absolute honour. I may just begin to make this my mission.

We all have a story in us. Most people find peace in

writing them too. Even if it's just for ourselves, to get it out of our head and onto paper. Being a part of that life journey is my inspiration to continue to do what I do with complete joy. Which, let's face it, is the meaning of life.

Ben

When Louise asked me to contribute to the anthology, I had to think more outside my comfort zone. What did I want to accomplish? By contributing, what do I want to accomplish? There were a lot of reasons why. Some were for therapeutic purposes – there's so much tension in the world and so much distrust, I needed an outlet.

But like many others, I wanted to share my story. One of my biggest fears is that after this pandemic is over, that everyone will try to return to normal. The way we treat each other will have changed. The way we do our shopping, the way we communicate with each other. The people we have lost. The sacrifices that have been made. And that's why I decided to write this.

If I can get one person to read our stories, and feel encouraged enough to share their own, then this won't be forgotten. All of us as artists, creatives, and people need to be there for each other and make sure that our stories are not forgotten. We must be there together for each other, and not let each other down.

We will survive. Together. One story at a time. So please, I beg you all. Write.

Nicola

Once we have our new normal and we're all able to go about our daily lives happily manoeuvring ourselves less than 2 metres apart and being able to clear our throats in public without fear of being flogged, our memories of lockdown will start to fade. The stories we tell future generations who listen wide eyed about those days when we fought over toilet rolls in supermarkets and half empty bottles of antibacterial hand wash could be bought on eBay for £20 will become laughable legends. That's why I wanted to contribute to this anthology and in doing so I feel like I have created my own, accurate little piece of the legend that will never be able to fade.

Joe

I am not a life writer, by advocation I write fantasy, I write sci-fi, I write games. So, writing for this book was way outside my comfort zone. But desperate times, desperate measures. When this lockdown came anxiety also locked down my creativity. No writing, no planning, just worry. When the opportunity to write for this book came up, I took it up almost out of desperation, a chance to write something, anything. I'm glad I did. Wrestling all those emotions on the page let me sort order out of chaos and make sense of what I was feeling. It helped. A few days after I finished the piece I started working on fiction again.

Ronnie

Ok, I was asked to write a piece for what writing has done for me, I will now endeavour to explain.

Nearly six years ago I became paralyzed. It wasn't a dramatic accident or anything quite that exciting. I've had spinal problems from childhood and after fifty years my luck ran out, I reached for the kettle and like a Marionette with its strings cut I slid to the floor never to walk again.

After a long while of adjustment, tears and tantrums, I had to look for other outlets, other ways to create and share my ideas.

Writing became the medium, I primarily write poetry on a variety of subjects.

Writing has allowed me to travel in my head to explore feelings and desires I hadn't taken any notice of while I was out and about living my "normal" life.

The Covid-19 lockdown is, for me, not much if a game-changer, I spend a large proportion of my time in bed due to pressure sores.

Writing is my window to the outside, it's my way of telling the world,

"hey I'm still here, I'm still fighting the good fight."

Without it, (without writing) I guess I'd be a silent statistic. A name on a district nurses call list.

As a parting word (and I hope this gets left in) I'd like to thank Lou for her support, her guidance and for the 'go for it' mentality she has instilled in me.

Following on from Pip's conversation with Louise, as detailed in this book, I wrote a poem for Louise:

Oh to be a Queen

You've been through some hard times, when life seems cold and bleak, when clouds are all around you and its Sunshine that you seek

When the tears of pain and heartache, stain your pretty face, you still get up to face the world, with courage and with grace

Many years you've been alone thinking love has passed you by, but no man can get passed your barriers no matter how they try.

You've built yourself a castle, with high walls and a keep, protecting a once broken heart, from those who try to peek.

But not all man are monsters, you might find a Knight, looking for his princess, to love and treat her right.

So pull up the portcullis, let your drawbridge down, give a Knight a fighting chance to win you love and crown.

Let him ride his noble steed right into your dream, let him be your true loves King and you could be his Queen

Give a Knight your favour, when he's jousting against the rest, once he's beaten all comers, you'll know his love is best

So if you want to be treated like the one and only Queen, tear down your barriers and start your true loves dream.

Lulu

As a child we are told not to verbalise a wish, for then it will never manifest. Then as adults we are offered contradictions: in order to facilitate something, we need to give it shape and form through words, spoken or written. Yet there's still the fear that once sounded or scripted we're giving these thoughts a tangible form which is present and therefore unquestionable. So, writing for me has always been a series of abstract words, notions and thoughts which sometimes, but not always offer a degree of coherency for a third person but ultimately is an outlet for my fears, loves, unspoken desires and dreams, a space to break rules and dabble with the taboo. A deliciously selfish journey of intellectual masturbation. So, to have written during these times is a bitter sweet experience. It's an acknowledgement forever that this is happening, now, for real.

The words on the page can now never be erased, nor these times in which we find ourselves. We are writing history. I haven't written for me, for over a decade. But in the face of the gargantuan unknown which we now face, there can be no grater fear than fear itself. So, writing this now has pulled the plug on all of those self-imposed restrictions. And the pouring forth of that stream, actually more of a swirling, cacophonous, and terrifying river of thoughts has been exhilarating and rejuvenating. Thank you for reminding me to live.

Jasmine

I'm not a writer but it was good to get creative and share my experience on this pandemic. I wanted people to understand how strange it was to feel afraid of other humans while I was at work.

Also, I think some people were very unhappy with the experience of being shut away but I was okay. It's nice to think that writing this piece might be a little slice of history to some future generations.

Tesco worker

I enjoyed writing this, it felt great. Getting the thoughts onto paper was a bit strange. It showed me how many other things came to light too while I just let my fingers do the talking on the keyboard.

I'm a little bit hooked now, I might write something else now!

Glossary

After this is over – a common term used during lockdown in reference to something which would happen post lockdown.

BoJo – an affectionate term used for Boris Johnson.

Cobra meetings – meeting rooms in the Cabinet Office in London. The rooms are used for committees which co-ordinate the actions of Government bodies in response to crisis.

Disease - a broad term used to refer to any condition that impairs the normal functioning of the body.

Epidemic – the rapid spread of disease to a large number of people in a given population in a short period of time.

Epidemiology – the study and analysis of the distribution (who, when, and where), patterns and determinants of health and disease conditions in defined populations.

Flatten the curve – Flattening the curve is a public health strategy introduced during the 2019–20 COVID-19 pandemic. The curve being flattened is the epidemic curve, a visual representation of the number of infected people needing health care over time.

Furloughed – a temporary leave of employees due to special needs of a company or employer, which may be due to economic conditions at the specific employer or in the economy as a whole.

Hazard boxes – social distancing measures that were put into place during the pandemic. Tape was stuck on the floor to tell people where to stand to keep a 2-meter distance.

Hot zone – an area likely to contain patients carrying the virus.

Isolation – various measures taken to prevent contagious diseases from being spread

It's okay not to be okay – a term used frequently during the pandemic in relation to people's mental health.

Lockdown – A **lockdown** is a protocol that usually prevents people from leaving an area. The protocol can usually only be initiated by someone in a position of <u>authority</u>. Lockdowns can also be used to protect people from a threat or other external event.

Loo Rolls – tissue used in the lavatory which was panic

bought during the 2020 pandemic.

Mortgage holidays – permission to miss mortgage payments while the pandemic lockdown was occurring.

New normal – a popular term used during the pandemic referring to the new life everyone adopted.

Pandemic – a disease epidemic which has spread across a large area, multiple continents or worldwide and affecting a substantial amount of people.

Quarantine –a restriction on the <u>movement of people</u> which is intended to prevent the spread of <u>disease</u>.

R rate – Basic reproduction number of an infection, the expected number of cases directly generated by one case (eg: Covid-19) where all individuals are susceptible to infection.

Rollercoaster – a description of the myriad of feelings experienced by the population as it changed on a day to day basis.

Self-isolate – a voluntary decision to stay away from others to prevent the spread of disease.

Social distancing – **Social distancing**, also called **physical**

distancing, is a set of non-pharmaceutical interventions or measures taken to prevent the spread of a contagious disease. maintaining a physical distance between people of two meters during the Covid-19 pandemic was thought to help prevent the spread of the virus.

The peak – when scientific data reached the maximum number of cases during the pandemic

Touch points – areas where humans touched and potentially left infectious microorganisms behind.

Twelve weeker – a name given to a person with an underlying health condition or susceptibility to a severe reaction to the Covid-19 virus. These patients were advised to remain in lockdown for a period of twelve weeks.

Unprecedented – never done or known before

Virology – the study of viruses

Zoom – video webinar software system widely used during the pandemic to prevent face to face contact through meetings. Often seen hosting virtual nights out, gym sessions and quizzes. Sometimes, according to Instagram, pink gin was involved. This is not scientifically proven.

Covid-19: How it made us feel

ABOUT THE AUTHOR

Louise Usher lives in Kent, UK with her twins and snoring shihtzu. Travelling widely to continue her writing while drawing inspiration from sandy shores, she films these trips for YouTube too. All while fighting to keep her heart from turning to ice.

A recent creative writing MA has seen her realise the dream of writing about life. Her current work in progress is her own painful auto ethnography of IVF and subsequent single parent heartbreak.

As a child she always had pen to paper, writing many letters to thankful Grandparents. The passion writing about life has taken Louise on a journey. Twenty years ago she wrote two novels and decided to leave them in her bedroom. Long before the days of typing on a computer, she also won a place in a publication called the Other Side and took her poem, 'Thank you, mother nature' to print. Reflecting on her own infertility issues, this painfully written poem was a total contrast to her now humorous journals that she writes and publishes.

www.louiseusher.co.uk

www.youtube.com/louiseusher

Covid-19: How it made us feel